Agroecological transitions, between determinist and open-ended visions

Peter Lang

Bruxelles · Bern · Berlin · New York · Oxford · Wien

Claire Lamine, Danièle Magda, Marta Rivera-Ferre,
Terry Marsden (eds.)

Agroecological transitions, between determinist and open-ended visions

EcoPolis
Vol. 37

Le Rayon Vert, photo issue de la série - L'Oiseau qui s'efface - ©Véronique Brill, veroniquebrill.fr

Financed by the INRAE SAD department (now ACT).

Library of Congress Cataloging-in-Publication Data
A CIP catalog record for this book has been applied for at the Library of Congress.

This publication has been peer reviewed.

Open Access: This work is licensed under a Creative Commons Attribution Non Commercial No Derivatives 4.0 unported license.
To view a copy of this license, visit https://creativecommons.org/licenses/by-nc-nd/4.0/

@ Claire Lamine, Danièle Magda, Marta Rivera-Ferre, Terry Marsden (eds.),
2021
1 avenue Maurice, B-1050 Bruxelles, Belgique
www.peterlang.com ; brussels@peterlang.com

ISSN 1377-7238
ISBN 978-2-8076-1852-7
ePDF 978-2-8076-1853-4
ePub 978-2-8076-1854-1
DOI 10.3726/b19053
D/2021/5678/24

Bibliographic Information published by the Deutsche Nationalbibliothek The Deutsche Nationalbibliothek lists this publication in the Deutsche Nationalbibliografie; detailed bibliographic data is available online at http://dnb.d-nb.de.

Table of contents

List of contributors 11

Acknowledgements 17

Foreword 19

Preface: Branching pathways in agroecological
transformations 23
    ANDY STIRLING

Introduction: Taking into account the ontological relationship
to change in agroecological transitions 33
    DANIÈLE MAGDA, CLAIRE LAMINE, TERRY MARSDEN,
    MARTA RIVERA-FERRE

Intertwining deterministic and open-ended perspectives in the
experimentation of agroecological production systems:
A challenge for agronomy researchers 57
    MIREILLE NAVARRETE, HÉLÈNE BRIVES, MAXIME CATALOGNA,
    AMÉLIE LEFÈVRE, SYLVAINE SIMON

Plant breeding for agroecology: A sociological analysis of the
co-creation of varieties and the collectives involved 79
    SOPHIE TABOURET, CLAIRE LAMINE, FRANÇOIS HOCHEREAU

Agroecological transitions at the scale of territorial agri-food
systems 101
    MARIANNE HUBEAU, MARTINA TUSCANO, FABIENNE BARATAUD,
    PATRIZIA PUGLIESE

How policy instruments may favour an articulation between
open ended and deterministic perspectives to support
agroecological transitions? Insights from a franco-brazilian
comparison 129
   CLAIRE LAMINE, CLAUDIA SCHMITT, JULIANO PALM,
   FLORIANE DERBEZ, PAULO PETERSEN

Teaching, training and learning for the agroecology transition:
A French-Brazilian perspective 153
   MOACIR DAROLT, JULIETTE ANGLADE, PASCALE MOITY-MAÏZI, CLAIRE
   LAMINE, FLORETTE RENGARD, VANESSA ICERI, AMÉLIE GENAY,
   CRISTIAN CELIS

The manufacture of futures and the agroecological transition.
Deciphering pathways for sustainability transition in France ... 177
   MARC BARBIER, SARAH LUMBROSO, JESSICA THOMAS, SÉBASTIEN TREYER

How access and dynamics in the use of territorial resources
shape agroecological transitions in crop-livestock systems:
Learnings and perspectives 199
   VINCENT THÉNARD, GILLES MARTEL, JEAN-PHILIPPE CHOISIS,
   TIMOTHÉE PETIT, SÉBASTIEN COUVREUR, OLIVIA FONTAINE,
   MARC MORAINE

The dynamics of agropastoral activities with regard to the
agroecological transition 225
   CHARLES-HENRI MOULIN, LAURA ETIENNE, MAGALI JOUVEN,
   JACQUES LASSEUR, MARTINE NAPOLÉONE,
   MARIE-ODILE NOZIÈRES-PETIT, ERIC VALL, ARIELLE VIDAL

What models of justice for the agroecological transition?
The normative backdrops of the transition 245
   PIERRE M. STASSART, ANTOINETTE M. DUMONT, CORENTIN HECQUET,
   STEPHANIE KLAEDTKE, CAMILLE LACOMBE, MATTHIEU DE NANTEUIL

Thinking through the lens of the other: Translocal
agroecology conversations 267
   DIVYA SHARMA AND BARBARA VAN DYCK

Table of contents

The rhetorics of agroecology: Positions, trajectories, strategies 289
MICHAEL BELL AND STÉPHANE BELLON

Postface 311

List of contributors

Juliette Anglade
Environmental scientist, INRAE, Aster, F-88500, Mirecourt, France

Fabienne Barataud
Geographer, INRAE, Aster, F-88500, Mirecourt, France

Marc Barbier
Sociologist, LISIS, Univ Gustave Eiffel, ESIEE Paris, CNRS, INRAE, F-77454 Marne-la-Vallée, France

Michael Bell
Sociologist, University of Wisconsin-Madison, USA

Stéphane Bellon
Agronomist, INRAE, Ecodéveloppement, F-84140, Avignon, France

Hélène Brives
Sociologist, ISARA, F-69007, Lyon, France

Maxime Catalogna
Agronomist, INRAE, Ecodéveloppement, F-84140, Avignon, and Department of Drôme, F-26000, Valence, France

Cristian Celis
Sociologist, INRAE, Ecodeveloppement, F-84140, Avignon, France

Jean-Philippe Choisis
Livestock scientist, INRAE, UMR SELMET, F-97410, St Pierre, France

Sébastien Couvreur
Livestock scientist, ESA, URSE, F-49007, Angers, France

Moacir Darolt
Agronomist, IDR-Paraná, Institute of Rural Development of Paraná, Curitiba, Brazil

Matthieu de Nanteuil
Sociologist, Université de Louvain CriDIS, IACCHOS, Belgique

Floriane Derbez
Sociologist, INRAE, Ecodeveloppement, Avignon, France

Antoinette M. Dumont
Agro-economist, SAW-B (Solidarité des Alternatives Wallonnes et Bruxelloises), Belgique

Laura Etienne
Pastoralist, Idele, UMT Pasto, Montpellier, France

Olivia Fontaine
Livestock scientist, CIRAD, UMR SELMET, F-97410, St Pierre, France

Amélie Genay
Agronomist, DRAAF Occitanie, Toulouse, France

Corentin Hecquet
Sociologist, Université de Liège, SEED, Belgique

François Hochereau
Sociologist, INRAE, SAD-APT, Paris, France

Marianne Hubeau
Bioscience engineering / agricultural economics, ILVO, Institute for Agricultural, Fisheries and Food Research, Merelbeke, Belgium

Vanessa Iceri
Geographer, AgroParisTech, Clermont-Ferrand, France

Magali Jouven
Livestock scientist, Montpellier SupAgro, UMR SELMET, Montpellier, France

List of contributors 13

Stephanie Klaedtke
Agronomist, Université de Liège, SEED, Belgique

Camille Lacombe
Agronomist, INRAE, UMR AGIR, F- 31326 Castanet- Tolosan, France

Claire Lamine
Sociologist, INRAE, Ecodéveloppement, F-84000 Avignon, France

Jacques Lasseur
Livestock scientist, INRAE, UMR SELMET, Montpellier, France

Amélie Lefèvre
INRAE, Agroecological vegetable systems Experimental Facility, F-6600, Alénya, France

Sarah Lumbroso
Agronomist, AScA, F-75010 Paris, France

Danièle Magda
Ecologist, INRAE, UMR AGIR, F-31326 Castanet-Tolosan, France

Terry Marsden
Geographer, Cardiff University, Sustainable Places Research Institute, Cardiff, CF10 38A, Wales, UK

Gilles Martel
Livestock scientist, INRAE, UMR BAGAP, F-49000, Angers, France

Pascale Moity-Maïzi
Anthropologist, UMR SENS, Institut Agro, Montpellier, France

Marc Moraine
Agronomist, INRAE, UMR INNOVATION, F-34060, Montpellier, France

Charles-Henri Moulin
Livestock scientist, Montpellier SupAgro, UMR SELMET, Montpellier, France

Martine Napoléone
Livestock scientist, INRAE, UMR SELMET, Montpellier, France

Mireille Navarrete
Agronomist, INRAE, Ecodéveloppement, F-84140, Avignon, France

Marie-Odile Nozières-Petit
Livestock scientist, INRAE, UMR SELMET, Montpellier, France

Juliano Palm
Sociologist, CPDA, UFRRJ, Rio de Janeiro, Brazil

Paulo Petersen
ASPTA, Rio de Janeiro, Brazil

Timothée Petit
Livestock scientist, ESA, URSE, F-49007, Angers, France

Patrizia Pugliese
Economist, CIHEAM Bari, Mediterranean Agronomic Institute of Bari, Italy

Florette Rengard
FAR, International Network for Agricultural and Rural Training, Montpellier, France

Marta Rivera-Ferre
Livestock scientist and sociologist, INGENIO (CSIC-Universitat Politècnica de València), Valencia, Spain

Claudia Schmitt
Sociologist, CPDA, UFRRJ, Rio de Janeiro, Brazil

Divya Sharma
Sociologist, Science Policy Research Unit, University of Sussex

Sylvaine Simon
Agronomist, INRAE, UERI Gotheron, F-26320, Saint-Marcel-lès-Valence, France

List of contributors 15

Pierre M. Stassart
Sociologist, Université de Liège, SEED, Belgique

Andy Stirling
Science and Technology Studies, Centre on Social Technological and Environmental Pathways to Sustainability (STEPS), Science Policy Research Unit, University of Sussex

Sophie Tabouret
Sociologist, CSI, i3, Mines ParisTech, PSL University, CNRS, Paris, France, INRAE, SAD-APT, Paris, France

Vincent Thénard
Livestock scientist, INRAE, UMR AGIR, F-31320, Castanet-Tolosan, France

Jessica Thomas
Sociologist, LISIS, Univ Gustave Eiffel, ESIEE Paris, CNRS, INRAE, F-77454 Marne-la-Vallée, France

Sébastien Treyer
Agronomist, IDDRI (Institut du Développement Durable et des Relations Internationales), F-75006 Paris, France

Martina Tuscano
Sociologist, INRAE, Ecodeveloppement, F-84140, Avignon, France

Eric Vall
Livestock scientist, CIRAD, UMR SELMET, Montpellier, France

Barbara Van Dyck
Political Agroecologist, Centre for Agroecology, Water and Resilience, Coventry University

Arielle Vidal
Livestock scientist, CIRAD, UMR SELMET, Montpellier, France

Acknowledgements

Thank you to all the participants in the founding seminar of this book, held in October 2018 at Le Pradel, France, to Laurent Hazard who helped us to convene it and Pedro Lopez Merino to conceive it, and to all the 51 contributors who built it up with the contributions of their chapters or through their introductory writings. We particularly thank Andy Stirling for the very inspiring preface he provided to this book.

Thanks to all the reviewers who provided critical and constructive comments to the different chapters: Colin Anderson, Marc Barbier, Philippe Baret, Gianluca Brunori, Jenny Calabrese, Benoit Dedieu, Les Levidow, Feliu Lopez Gelats, David Meek, Jean-Marc Meynard, Ana Moragues, Manuel Gonzalez de Molina Navarro, Erin Silva, and to the anonymous reviewers of Peter Lang's editorial committee.

Thanks to Xavier Arnauld de Sartre and Olivier Petit who hosted and followed this project in their Ecopolis collection.

Thanks to the INRAE SAD department (now ACT) that financed this publishing project.

Foreword

by Philippe Mauguin

President and chief executive officer of INRAE

By the end of 2020, the idea that the Covid-19 crisis was a mere parenthesis had dissipated. By measuring the impacts on our personal, social, cultural and economic lives, we all began to perceive that there was a before and an after, in other words that we had found ourselves in a deeper sort of transition.

It is as if the entire notion of "transition" has moved beyond the scientific sphere to apply to society as a whole. This new health situation thus makes this book all the more relevant. Of course, the aim here is to study the processes of a particular transition – the agroecological transition – but the questions raised by the group of 51 researchers from eight different countries will certainly find echoes in the debates within the political world and civil society: How can we understand and describe the processes of change? Must we have a specific goal for the transition – the "after" – already in mind, or can it be developed throughout the transition process itself with some room left for indetermination?

This research and work on the agroecological transition are especially necessary as it is a complex phenomenon. It is complex first of all because it is multidimensional: the changes are technical, social, ecological and political all at the same time. But it is also complex because it takes place at several scales, from the farm (or even the farm plot) to the food systems.

Given such complexity, it is useful to delve into the "hows" of the agroecological transition. Debates are thus open among scientists regarding the processes of change and their different visions of change. Should a global and systemic approach be favoured – one that involves more than a particular profession or economic sector – since the transition is not strictly limited to farmers but requires the mobilization of the entire chain, from production to consumption? Or would it be better to take an approach based primarily on the drivers of technical innovation? Is it necessary to adopt a "deterministic" perspective based on the predetermination of the end goal or an "open-ended" perspective that views indetermination as an asset?

By informing this debate and taking account of the many ways the agroecological transition can be addressed, this book aims to make visible the "state of the art," to give tools to those implied in agroecology and to contribute to the expanding field of research on transitions.

At INRA, this subject was handled by the Science for Action and Development (SAD) Division, which became since 1 January 2020 within INRAE the Sciences for Action, Transitions and Territoiries (ACT) Division. This division supports an approach to agroecology that takes actors' visions and practices into consideration. In this way, it has helped to formulate research questions about the transition, focusing on processes of change and not just the application of agroecological principles. This book is a product of the division's scientific priority "agroecology for action." This is why it includes contributions from not only INRAE researchers but also scientists from french and foreign universities. This is also why it adopts a resolutely interdisciplinary approach by drawing from the life sciences, technical sciences and social sciences.

Each disciplinary field represented in the book sheds light on the connections between the "deterministic" and "open-ended" perspectives and analyzes how "deterministic" phases and "open-ended" phases can alternate. In the end, it is indeed the combination of these trajectories that should enable the large-scale deployment of agroecology.

Without concluding or settling the debate, this book provides arguments in favour of transition paths where actors are stakeholders who are involved in governance arrangements, and where scientific knowledge and practical knowledge are hybridized within new modes of collaboration with all public and private actors, within spaces such as living labs, transformative labs and multi-actor observatories.

Making this work visible means opening up avenues for the development of approaches for the design, innovation and support of the agroecological transition and of public policy instruments, while taking into account both deterministic and open-ended visions that will guide this transition. It also means questioning our relationship to change. Finally, it means understanding how agroecology can be a framework to integrate and drive forward the study and design of more sustainable agricultural and food systems.

Preface
Branching pathways in agroecological transformations

Professor Andy Stirling

In what other area of life, could experience of transformation be more profound or intimate than in the food we eat? Revolutions in life-ways around food occur at some of the deepest, most pervasive and most momentously formative levels of culture, identity and history.

Provisioning of food not only implicates a multitude of understandings, practices and institutions, but also constitutes them. And what is also formed, reshaped, or exterminated in the process are not just routine habits and structures around food and its production – but (over time) entire cultures, geographies, and species. Whether in physiological, societal or the broadest of evolutionary terms, we (as a world) are – quite literally – what we eat.

This much hardly needs rehearsing in a preface to a largely Francophone book – least of all in an English-speaking voice! In few places beyond France is this societal centrality of food more clear. And where more than in France have government and academia alike (to their credit) taken duly seriously the imperatives to reverse the recent trend, in which the making of food has moved from an affirmation of culture to a planetary threat? For among the many potentially existential challenges of the contemporary world, there is none in which food production is not central.

Currently globalizing industrial ways to grow, manufacture and distribute food are among the strongest drivers of climate disruption. The global spread of agrochemicals is seriously exacerbating wider air and water pollution. Directly and indirectly, pesticides take a terrible toll on biodiversity, threatening extinction to many vital pollinators. Forests are levelled for intensive grazing, heavy ploughing destroys soil structures, and uprooting hedgerows helps wash vital earth away. Applications of nutrients, herbicides and monocultures blight entire ecosystems. Chemical spray drifts and residues combine with food additives and intensive processing to compromise human health.

In associated growth of inequality and stratification around the world, status anxieties help drive shifts towards more meat eating, amplifying environmental impacts. In the resulting expanding and concentrating of food infrastructures, peasant farmers and agricultural workers become ever more excluded and oppressed – alienated from identity-defining, community-sustaining livelihoods. In consumption and the home, as much as the workplace, mechanization and marketization of food practices add to the anomies of Modernity.

Amidst all this unruly, distributed and emergent complexity, there are of course some highly active interests at work. Encompassing landholding, data processing, resource provision, supply chains, intellectual property, retail channels, and global webs of value, transnational corporations entrench their incumbent privilege and concentrate their patterns of appropriation. Whether on labour protection, food safety or agricultural trade, worldwide regulation is often better seen as a captive aid to these interests than as a counterbalancing force.

Crucially, however, it would be a mistake to see behind all this, some singular sentient controlling "*invisible hand*." The unfolding of global food systems is not like a deterministic machine, responding fully and without side-effects, to the pressing of corporate buttons or the pulling of policy levers. The picture is at least as much one of unintended impacts and collateral effects as of any unitary orchestrating agency. Although externalized as much as possible, the many manifest risks and disruptions also come back in the longer run, to threaten those interests which initially benefit.

So whilst unfolding directions of change are clearly influenced in profound ways by incumbent interests, these are important less as conspiracies and more in the shaping of broad gradients for change. Imaginations of machine-like control are crucial, but – as has always been the case in Modernity – these are not about reality as it is, but about the reality it is expedient to portray. Incumbent interests do not solely "make the waves" of change, they are more concerned with surfing them. It is through stories of control, that elite agents who are actually *unable* to control the world in any full (machine-like) sense, are nonetheless best able accrue the benefits of continuing privilege.

It is these dynamics of privilege (rather than control) that appropriate, concentrate and accumulate a continuously growing and ever less equitable share of value from worldwide food systems and their associated flows. But these benefits are persistently at the mercy of unintended and unexpected "events." In seeking to transform global food practices, it would therefore be a grave mistake to assume there is some lofty citadel of agency, which must somehow be seized... still less petitioned. In agriculture and food (as in every other major area of activity in the world), it can be a truth that is as uncomfortable for radicals as for conservatives, that *there is no cockpit*.

But modernistic imaginations of control are nonetheless ever more overbearing in their impacts on global agriculture and food. Arguably the best example, lies in accelerating expectations over recent years, around the manipulation of living genomes. As breathless storylines burgeon around this engineering of the very stuff of life, new forms of uncertainty, irreversibility and oppression intensify. Potent as it is, this reinforces the fallacies, fictions and fantasies of control that are already helping to drive the other challenges around food.

In plant and animal gene editing, as in satellite-assisted farming, as in synthetic meats, as in just-in-time supply chains, "precision" is confused with accuracy and "data" with meaning. Computational methods, mechanistic metaphors and narratives of control continue to proliferate across agriculture and food, recasting the living world and its products in an ever more instrumental idiom.

So, despite its many manifest failings in unfolding real-world events around agriculture and food, the controlling imagination of Modernity is doubling-down in a time of crisis. And despite their neglected flaws, intensified notions of precision successfully consolidate essentially colonial processes of appropriation and concentration (which actually depend less on exactitude than on main force). Although the world remains uncontrolled, waves of change are surfed to make the diverse and dynamic complexities of subjective life-ways around food ever more reduced to objectified extractable, consumable resources.

So, even where visions of control are manifestly falsified by unintended impacts on ecologies and on people, the underlying engineering storyline asserts ever more strongly, the controlling aspirations of Modernity. The effect is to emphasize in collective imaginations, an excluding focus on the aims and actions of industrial and technological interests – eroding other forms and sources of agency – among farmers, rural communities, other beings, individual ecosystems, evolution and Nature as a whole. All, after all, are engaged in their own agency-defining processes of *'orienting among alternative pathways for change'*. Seeking to domestic Earth's other richly-nested agencies, Modernity solipsistically imposes its own narrow way of being, as yet another self-destructive monoculture.

As this book so ably documents, this is the tangle of intractable challenges to which the equally multiple responses called *"agroecology"* have arisen. Led by a global grassroots agroecology movement, various streams of broadly related activity include: struggles to uphold many hitherto suppressed indigenous and vernacular knowledges and practices; specific cultivation techniques like multi- and inter-cropping, companion planting or biological pest control; low-input approaches like organic farming, biodynamics and integrated pest management; organizational innovations like open source seed sharing, micro-finance and self-organized extension services; cultural shifts like veganism, vegetarianism or buying-local; consumer initiatives around fair trade, slow food and labelling; institutional transformations around box schemes, community farming and urban gardening; political reforms towards co-operative networks, price support and women's farming initiatives; mobilisations to defend smallholder livelihoods, labour rights, environmental standards and consumer protection as well as more radical collective action towards land and income redistribution, worker and rural autonomy and many other kinds of emancipation.

For sure, this potentially paralyzing plurality, scope and complexity of agroecology, makes it more clearly motivating to frame action simply and positively rather than negatively and ambiguously. Choosing a convenient single label for action, can be central to galvanizing momentum. But right at the outset a key query arises. Why given all the disparate (and sometimes strongly contending) aspects of agroecology, is it becoming so widespread to use such a deceptively *singular* term? Does this apparent unity act more to help or hinder transformations to more ecological and socially just food production? If the problem lies in homogenizing agency, what is the price of this kind of standardizing in resistance?

Of course, if what is sought is patronage from incumbent interests – in science-driven, technologically-enabled change, powered by top-down visions – then this kind of *"keep it simple"* singularity can be highly expedient and tactically effective. For this purpose, aspects of agroecology can be pursued through the lens of fear-driven (rather than hope-inspired) technical agendas around a supposedly unitary *"nexus"* of *"grand challenges."*

With urgency supposedly brooking no time for diversity or deliberation, this is too often the style of *"transition management."* Here, with assertive singularity typically emphasized by the definite article, visions of *"the"* sustainable food transition can offer a tempting rallying banner – especially near to 'the corridors of power'. But tactical pressures and academic incentives for "policy impact" can wreak real damage to transformative strategic ambitions.

In a kind of "Stockholm syndrome," for instance, particular strands of agroecological practice can find themselves captured by the modernistic controlling idiom they ostensibly oppose. Under seductive pressures for incumbent patronage, ecological and political transformations can find themselves restricted to mere modulations of continuing productionist imaginations of relations between society and food. Cosmetic accretions of green-tinged agroecological *"niches"* around the margins, can help conceal and protect persistent business-as-usual.

This can occur, for instance, where supposedly alternative agroecological practices find themselves advocated or implemented in ways that uphold the same tropes of control that drive many of the food-related problems of Modernity in the first place. If these practices are framed primarily as *"innovations"* to be *"scaled up"* and *"disseminated"* then the dangers are real. Even if only inadvertently, such *"mainstreaming"* of agroecology can reproduce the same pattern of expert-led *"evidence based policy,"* competitively pursuing *"pro-technology"* strategies for *"world beating"* *"solutions"* that are (if inadvertently) actually driven by far narrower interests.

Albeit with different vocabularies and focusing on contrasting techniques, then, this style of agroecological transition, can still end up supporting the same processes of appropriation and concentration around essentially the same globally-stratified structures of patronage and privilege that are shaping the problems in the first place. For instance, by treating action as depending on knowledge, more than the other way around, artisan practices, vernacular knowledges and entire indigenous worlds, still get bulldozed by Modernity.

This is worth considering hard, for there may be much more at stake here than irony. What if there is at least some sense in which *"the medium is the message"* in political change around agriculture and food? Rather than depending on the reductive – domesticating and standardizing – etiquettes of "scaling up," what if progressive change is a more dynamic and holistic process of resonance and prefiguring in the vibrant "thriving chimes" of persistently contrasting practices? What if pluralities in how processes of agroecological transformation unfold, matter as much as the diversities with which they supposedly end?

Is it not a lesson of the past half century of struggle in the food sector, after all, that research and innovation can be essential, but are necessary rather than sufficient. Science and technology are always better seen as tools, than as ends in their own right. There are many ways and means to enhance environmental efficiency and reduce waste, for instance, but – as with all kinds of efficiency – these always depend on the choice of denominator. There is no kind of efficiency that forms an end in itself. So, might it not be precisely when *instruments* become confused with *aims* of transformation, that the process becomes vulnerable to manipulation by existing incumbent interests simply donning new clothes? If so, there seems little reason why agroecology should be uniquely immune?

To be more specific: if the roots of so many ecological devastations and social injustices lie in fictions, fallacies and fantasies of control, then is it not counter-productive simply to clothe these underlying syndromes in ostensibly greener forms? Might not agroecology be as much about new *processes* of social political change, as novel technological *endpoints*?

If this cautionary note holds true, then an overly abstract and supposedly unitary notion of agroecology might – despite the tactical temptations – become a self-defeating strategy. If agroecology itself is framed in an instrumental, scientific, pragmatically *"solutions-oriented"* way, might the forms it takes simply reproduce the structural forces it notionally resists?

All-too-easily, a conventional controlling notion of *"the* transition" to agroecology can obscure alternative visions not only of the many different dimensions and perspectives that this term encompasses, but of the underlying imaginations of what politics and society are all about. Efforts to promote agroecology by producing 'dashboards' for policy cockpits, are vulnerable to the perennial (but neglected) surprise that there is in fact no cockpit. Albeit in unintended ways, such vain efforts can suppress the messier and more unruly kinds of value, action and struggle on which progressive transformations have always arguably depended.

Instead of being science-driven in a circumscribed sense, then, agroecology might be seen to harness many other kinds of knowledge and practice – not just those that happen to be most privileged in academic disciplinary hierarchies. Research can still play essential roles, but the scope becomes wider and more socially open – and the power relations shift. Beyond narrowly restricted forms of material or biological engineering, a multiplicity of wider forms of innovation can flourish – not just technological, but behavioural, organizational, institutional, political and cultural.

And when it comes to identifying the particular directions in which such transformations should be oriented, then the role of civil society can come especially to the fore. Although often celebrated in retrospect by government, business and academia in the present, many of what can looking back, be seen as key steps in advancing agroecology – for instance, like low-input farming – were actually actively deprecated by these interests at the time.

Likewise, when social movements first drew critical attention to so many of the grim realities lurking behind loudly proclaimed promises of industrializing agriculture, they were at first very strongly resisted by precisely the kinds of actors in government, business and academia, who now loudly enroll these same problems as imperatives in their own "grand challenges". Sometimes the resulting platforms can seem more about the performing of grandness, than the critical imperative of challenge – distracting necessary action, more than reinforcing it.

So, without collective action in wider society – on indigenous rights, land tenure, biodiversity conservation, labour conditions, water quality, animal welfare, consumer standards – it is clear from issue after issue in the food sector (as elsewhere) that none of the great estates of governance could have been relied upon to lead the struggle against incumbent interests. Although key actors in business, government, the media and academia each variously fulfiled particular crucial roles in every stage, all typically required constant motivation and pressure from wider political mobilization in order to play their parts.

Inspired by values forged in subaltern solidarities, underpinned by knowledges born in generations of practice, driven by engines of collective action, steered by compasses of democratic struggle, pivoted in responsive interventions in academia and public agencies, nurtured by pioneering markets, and harnessing the distributed creativity and innovation of social entrepreneurship – it is more from messy politics and cultural expression than from neatly-scaled layers of patronage, that progressive transformations thrive.

For, looked at most generally, is it not exactly from these deeper, wider and more unruly roots, that related continuing progressive struggles have unfolded, like those against slavery, colonization, worker exploitation, class oppression, sexism, racism and homophobia? Policy, technological and scientific interventions have all played many vital roles. But the pace and direction of travel have flowed mainly with the deeper and more pervasive tides of political culture. There seems little reason to speculate that emerging agroecologies should be any different.

So, if it is the broad aim of agroecology movements, that food can be produced in ways that *care* for people and nature (rather than attempt to *control* them), then is it not a betrayal of possible forms of transformation, to seek to unfold them in the same old controlling ways? Is it not here again, that more plural, mutualistic, bottom up, hope-inspired, complexity-affirming and caring dynamics of *transformation* can show sharp contrasts with more competitive, hierarchical, fear-driven, simplistic and controlling disciplines of *transition*?

The aim here is – of course – neither to insist on particular words, nor to seek to police their use. Terms like "transition" and "transformation" can (as is evident in this book) be fairly used in many different ways. Nor is the implication that either of the ideal-types named this way above, can exist in any complete or isolated form. The point is rather that real-world struggles for change can be enacted in relational ways that variously approach only some among the multiple possible poles. How agroecology is pursued, can matter for how it turns out.

So it is around these key themes of diagnosis and prescription, that this book offers such a rich and diverse source of insight, experience, and information. The scope and diversity of associated issues is admirably reflected in the depth and richness of the contributions. The subtleties are well addressed not just by the nuanced style of discussion throughout – but also by the pluralities (and occasional constructive tensions) – between the contrasting perspectives formatively characterized in the introduction: "determinist" and "open-ended."

With repercussions unfolded in multiple ways throughout so many engaging chapters, it is this compelling general heuristic distinction between "*determinist*" and "*open-ended*" approaches to constituting and supporting agroecology, that forms a major contribution of this book. With complexity and uncertainty highlighted throughout, this is the main lens through which many crucial detailed implications for action and knowledge can be explored. It is here that contrasts and tensions are most generative, between what are described above as "caring" and "controlling" approaches – equally to agroecologies as end points and to the myriad associated dynamics of transformation themselves. As much in choices between determinist and open-ended means as in their contrasting intended ends, mediums can be messages.

Drawing on an unparalleled level and quality of experience across a diverse community of leading researchers and practitioners in different branches of agroecology, then, this book admirably resists temptations to assert "*one good vision.*" Instead of focusing on a supposedly singular agroecological transition, attention is given instead to the "*grounded relationships of change*" that make up the processes of transformation themselves.

Nor is agroecology itself cast in terms of any particular "*bounded or narrow 'technology',*" but instead as a plurality of wider innovations, practices, institutions and movements. In each respect, the book also benefits from drawing centrally on the experience of France – as a country, which (as this essay noted at the outset) has grappled more than is usual, with the challenges and opportunities involved. There is much that the unsurpassed quality of tradition, critical thought, experimentation and creativity undertaken in France, can offer to research and struggle in other settings. Engaging exactly with this process of dialogue across so many axes of difference, the book ably sets an agenda for onward learning on all sides.

So we return to the nature of the agroecological challenges with which this essay began. It is clear from all the many dimensions of current crises in global agriculture – each in itself a profound and intractable challenge – that a multiplicity of transformations are sorely needed. Agroecologies are not one thing, but many. There are multiple branching pathways through which to explore possibilities – and realize pluralities of potentialities. But as a focus both for analysis and mobilization, this multiplicity makes the overall issue all the more important. In seeking to grapple with – and do justice to – the diverse implications, there could hardly be a better spur, or guide, or framework, than what is provided in this book.

Taking into account the ontological relationship to change in agroecological transitions

Danièle Magda, Claire Lamine,
Terry Marsden, Marta Rivera-Ferre

1. Introduction

The intensification of researchers' work on agroecological transitions (AET) in recent years (Duru et al., 2015; Elzen et al., 2017; Levidow, 2015; Meek, 2016; Ollivier et al., 2018) highlights the growing importance of how change is addressed in the field of agroecology. This research offers a wide variety of interpretations, from different standpoints, on the most important mechanisms, knowledge, dimensions, and scales to consider for understanding *the change* towards agroecology and to support or launch it. Generally, these standpoints are based on a framing of what needs to be changed (and how) but not on the explanation of the *vision of change* they carry.

Yet, these interpretative choices are deeply embedded in what we may call our *ontological relationship to change*. This relation to change is obviously complex and multidimensional, but deeply framed by the way we face complexity, uncertainty, and radicality. When dealing with concrete transition processes, this relation to change is the foundation of the interpretations and proposals of the various involved stakeholders (researchers, farmers, advisors, facilitators, decision-makers) upon mechanisms of change and modalities for conducting transitions. They guide decisions on why, how and by whom changes are, or should be, enacted and implemented. However, these ontological relationships to change are seldom analyzed in studies dealing with transitions.

We posit that this aspect is important to acknowledge and characterize the emerging diversity of agroecological transition dynamics and projects. It therefore seems important to critically expose and clarify these ontological relationships, while instead the debates on *agroecological transitions* are generally obscured or confounded with controversies over definitions and visions of *agroecology* itself. This is largely explained by the fact that the recent literature on agroecology has mostly focused on these different *definitions* and *visions of agroecology* (Méndez et al., 2013; Norder et al., 2016; Rivera-Ferre, 2018). The debates and controversies surrounding these differences have either relegated the question of our relationship to change to a background factor, or simply ignored it. Definitions and visions of agroecology are most often discussed in a normative way and through oppositions on the role of stakeholders, the place of technology, the levels of organization, and the dimensions and scales taken into account. Dualistic oppositions such as weak versus strong (Duru et al., 2015; López-i-Gelats et al., 2016), soft versus deep (Dalgaard et al., 2003), technical versus political (Molina, 2013), co-opted (by corporate actors or governements) versus peasant or social movements' agroecology (Rivera-Ferre, 2018), or incremental versus radical agroecological transition (Berthet et al., 2016), dominate this literature. Agroecology is indeed the locus of vivid controversies arising along its process of institutionalization in many national contexts, and a "territory in dispute" (Giraldo and Rosset, 2018; Lamine, 2017). Whilst the question of (desirable) change is clearly central to these controversies, their focus is usually on the depth and radicality of the change proposed by a given vision, which tends to eclipse the ontological relationships of actors (or researchers) to the very "change process" itself.

This collective book is an endeavour to take into account and critically expose existing and grounded relationships to change, based on the idea that they constitute a fundamental and often underestimated background of the design of agroecological transition. Proposals to define ontological relationships to change can be obviously multiple (we could for example refer to different conceptions of time, that is, linear versus cyclical, or to different conceptions of the role of action or inaction in change, as in chinese philosophy (Julien, 2004)). In this book, we have chosen to refer to two ontologies; respectively *the deterministic and the open-ended* ontologies, as they represent two contrasting conceptions or interpretations of the change process, based on whether objectives and means, targets and pathways are predetermined or defined along the change process. In that way, they represent two ideal-typical stances when facing complexity and uncertainty. Determinism may be considered as the main framework of (our) thinking in western societies since

18th century. The use of the term "open-ended" to refer to an alternative ontological relationship to change is more recent and has taken off with sustainable issues, as we will see below.

Taking into account the fact that different visions about change are at play in transition processes, the issue is not to relate *a priori* one ontology to a vision of agroecology or to a category of actors, but to analyse through a comprehensive and non-normative approach how these visions are involved in the way actors (including researchers) deal with AET processes. Nor is our objective to argue that there is *one* good vision of AET, which should relate to an open-ended perspective. Instead, we seek to show that we need to cope with the difficulty that is generated by this tension between a deterministic and "classical" western vision of change and the acknowledgement of a certain indeterminism.

A key hypothesis here is that even if the open-ended ontology seems relevant facing sustainability issues, in reality the reference to/anchorage in one ontology or another occurs and operates in many ways, varying over time and scales and in different situations (both at the individual and collective levels). Yet, we still lack epistemological tools to analyse and take advantage of this diversity of combinations of ontologies for sustainability's stake. The analysis of these combinations is a major challenge both for those who analyse and those who implement agroecological transitions.

In this introductory chapter, we first examine the main characteristics that distinguish these two contrasting perspectives, to then analyse the ways in which they are present and discussed in the main theoretical currents about transitions towards sustainability (in the broader sense). Then we propose an interpretation of these two perspectives in the case of agroecological transitions. Finally, we introduce the collective writing process and the chapters of the book, according to the way the diverse authors discuss these perspectives – or suggest other ways to discuss relationship to change in the agroecological transition.

2. Characterizing determinist and open-ended perspectives for transition

Following its most common definition, determinism is a theory according to which events and the succession of phenomena are explained by a principle of causality (exposed already in Ancient Greece by Aristotle and the stoics). Under the same conditions, the same causes should produce the same effects. In this sense, determinism is joined with predictability. When a *future* transition is in question, the **interpretation of the mechanisms** of past change is the basis of the identification of mechanisms to favour in the conduct of future change. In this perspective, the **initial state and "closure" of a system** determine the field of causal possibilities. The system is defined here by the interdependencies of objects, processes, and actors involved in change. The change process is thus causally contingent on the system and its initial state. This determinist perspective predefines not only the **goal** (for example, a more sustainable agriculture) but usually the specific **target** with more or less precision (for example, 10% of organic farming and/or 50% in pesticide reduction), and very often, the possible **pathways** and **steps** (see Tab. 1). In this perspective, **complexity** is to be reduced to a limited number of analyzable blocks and variables, and **uncertainty** by knowledge production. In contemporary western societies, this ontology of change, largely irrigated by the development of sciences and experimental approaches, still deeply grounds the visions of transitions, although it most often remains implicit.

We do not ignore that various scientific fields anchored in deterministic perspectives have refined this archetypical vision over time. Within the physical sciences, for example, the quantum revolution has generated a detachment from the classical view of causality. The impossibility of direct measurements of some elementary particles' behaviour led to the development of a statistical conception of reality. This, and the consequent experimental un-verifiability, led modern physics to partly break up with the causality principle to understand and act upon the real world. In another way, the chaos theory recognizes that, due to their complexity and sensitivity, living systems can evolve in substantially different ways after slight deviations in starting conditions and without us being able to foresee it. It needs to be noted, however, that in this theory the perspective remains a mainly deterministic one, and the practical inability to have a perfect knowledge of the parameters is to blame for prediction errors. Thus, while this theory can concede that complex systems may partly defy prediction, it is nevertheless associated with an intention of reducing uncertainty – an uncertainty seen most often as lack of full knowledge of systems that could be ameliorated by better measuring instruments and statistical tools. This leads to theories assuming unpredictability as a fact while keeping a deterministic vision, as is the case with the determinist chaos theory.

In contrast to this determinist perspective of change, the "open ended" perspective posits that the **uncertainty** and **complexity** of mechanisms of change preclude any mechanistic and causal interpretation, thus preventing perfect objectivity. This is what Karl Popper, for example, proposed in his epistemological and ethical discussion of "indeterminism" (Popper, 1982). The interpretation of change mechanisms acknowledges uncertainty, complexity and subjectivity. Although a general **goal** (such as a more sustainable agriculture) can be predefined, as in the determinist perspective, it is not feasible to define *ex ante* precise **targets** to be achieved and **pathways** to be taken with possible intermediate **steps** clearly identified (Tab. 1). The reference to an open-ended perspective appears in the recent literature on sustainable transitions, defined by some authors as « continuous open-ended processes of societal innovation » (Loorbach et al., 2011). It also appears in critical comments on sustainable transitions approaches, suggesting that the notion of transformation is better suited than that of transition to the stakes that come with the type of radical and profound changes that our societies must implement to achieve sustainability, as we will see below (Marsden, 2017; Spaagaren et al., 2012; Stirling, 2014;). In a variety of approaches that have emerged around the issue of sustainable transitions, indeterminacy appears as an irreducible feature of the system – something to be dealt with or even to be regarded as an asset. In this open-ended perspective, and in contrast to the determinist one, the **system state** is considered as being evolutionary, context-dependent, and place-based. The closure and the structure of the system might also evolve along with the transition process. Whilst initial choices outline the system (and thus frame the initial issue as well as the stakeholders involved), this definition of the system and the targets and pathways can be redefined to varying degrees during the change process. This perspective, primarily defined in opposition to the previous one, can assume several forms or degrees, depending on whether its opening claims concern the target, the pathway and steps, or the interpretations of change mechanisms. In this open-ended perspective, and in contrast to the determinist perspective where **management styles** are based on steering and control, managing the transition is characterized by an on-going definition and redefinition of targets and pathways in the course of action, stemming from how the different actors interpret and experience the dynamics of the system. Moreover, in this open-ended perspective the **diversity and contestability of actors' visions** is to be acknowledged, clarified, and integrated, whereas it tends to be reduced through cognitive processes and/or deliberation in the determinist one.

In Tab. 1 above, we suggest an ideal-typical characterization of the two perspectives of change stemming from the two ontologies which are at play within change processes. This characterization aims at giving insights on how these perspectives frame options for interpretation and management of change process.

Tab. 1: A proposition to characterize the determinist and open-ended perspectives in the interpretation and the management of change process.

	Determinist perspective	Open-ended perspective
Interpretation of change mechanisms	Based on causality, considered as external and objective	Acknowledges subjectivity, complexity, and uncertainty
Role of the system state	Dependency on initial system state	System state considered as evolutionary, context-dependent, and place-based.
Goal	Defined	Defined
Target	Pre-defined with more or less precision	Not pre-defined
Pathway and Steps	Usually pre-defined	Not pre-defined
Complexity	To be reduced to a limited number of analyzable blocks	To be acknowledged within a holistic vision
Uncertainties	Considered as risks and to be reduced by knowledge production	Considered as intrinsic (may even be an opportunity)
Management styles	Steering to reach aim and target, based on control and planning and on pre-established knowledge of mechanisms and processes, as well as prediction tools	Managing the definition and redefinition of target and pathway during the course of action, based on the observation of and/or experimentation with mechanisms and processes
Diversity of visions	To be reduced by cognitive processes and/or deliberation	To be acknowledged, clarified, and integrated

The ontological relationship to change

We previously stated that the reliance on one or the other perspective is rather implicit. However, in many studies on agroecological transitions, non-determinist or open-ended *perspectives on transition processes* are often favoured (though sometimes expressed in other words), generally in line with a participatory approach. These are also strongly associated with more social or political *visions of agroecology* (Francis et al., 2009; Méndez et al., 2013; Molina, 2013). Conversely, visions of agroecology that involve technical or technological innovations are often likened to a command-and-control perspective, the implementation of primarily technological solutions, and optimization based on predetermined transition paths. An alignment between a vision of social agroecology and an open-ended perspective of change on the one hand, and between a techno-centric vision of agroecology and a determinist perspective of change on the other one, has become implicit in the discourses and debates over AETs. This has resulted in a lack of attention to the (different) visions of change themselves. Most debates on agroecological transitions are thus readily categorized in these dichotomies, often reinforced by tensions between disciplinary or research stances, which generally leads to undermining the real complexity and diversity of the relationships to change.

Different conceptual frameworks dealing with transitions towards sustainability (including in other fields than agriculture and food) have influenced, and can enrich, the thinking on change processes in AET. Understanding how these frameworks' ontologies echo with determinist or open-ended perspectives is therefore a necessary first step in the clarification of how these two perspectives may be combined in change processes in the case of AET.

3. How different conceptualizations of the transition to sustainability encapsulate determinist and/or open-ended perspectives on change processes

Beyond the specific case of the AET, how do the main approaches at play on transitions towards sustainability fit into one or the other of these two perspectives? Various theoretical frameworks have been developed over the last decades in the social sciences and in the epistemic communities that emerged around socio-ecological approaches, sustainability sciences and/or transition studies to address complex system changes. We will focus here on a few central ones that have inspired many recent approaches to agroecological transitions. The aim is not to re-produce a detailed analysis of each different theoretical framework's specific paradigms and ontologies but, more modestly, to identify their links to one or the other of these two ontologies – even when they are not always mentioned explicitly.

Critiques of determinism have strongly impacted the debates over technological change. For example, evolutionary economics has emerged based on criticism of single-factor determinism and therefore of classical theories focused either on the market (demand-pull theories) or on technological innovation alone (technology-push theories). Critics argue that technological trajectories are characterized by interactions between scientific innovations, economic factors, and institutional variables (and not by one-dimensional causality), and thus generate powerful exclusionary effects (Dosi 1982)[1]. Due to sensitivity to initial conditions and increasing returns, technologies with similar performance and functions and perhaps stronger long-term potential, are put aside (as exemplified by the well-known cases of Qwerty keyboards and of video formats; for a synthesis see Liebowitz and Margolis 1995)[2]. The concepts of path dependency and lock-in were proposed to explain how certain technologies prevail and acquire stability over time.

These approaches have partly inspired the "multi-level perspective" for change (MLP), according to which transitions result from interactions at three levels: landscape, regime, and niches (Geels, 2002). This approach sees niche innovations and changes in the "socio-technical landscape"[3] as a source of pressure on the (dominant) regime that generates destabilizing effects on it, and opens windows of opportunity for niche innovations. This approach was refined over time, and two contrasting processes of changes in rules and norms have been outlined: an evolutionary-economic process, which results from indirect market-related changes, and a social-institutional process, which results from

---

[1] See Possas et al. (1996) for an adaptation to the particular case of agriculture.

[2] Cowan and Gunby (1996) adapted these concepts to agriculture and more specifically to pesticides, and analyzed the impossibility of switching from chemical pesticides to Integrated Pest Management (considered as a competing technology) by showing how the combination of many factors creates a lock-in situation (Cowan and Gunby, 1996). Also see (Wilson and Tisdell, 2001; Vanloqueren and Baret, 2008).

[3] A term we borrow from (Rip and Kemp, 1998).

negotiations between actors (Geels and Schot, 2007). The same authors elaborated a typology of "transition pathways" based on two complementary criteria. First, the timing of interactions, along with notions of coincidence or time lags (for example, the fact that there are or not pressures from the "landscape" when there are relevant innovations under development in the "niches"). Second, the nature of interactions: whether interactions reinforce an evolution or cause a break.

The multi-level perspective is first and foremost an analytical framework that proposes an interpretation *ex post* of transition processes and more specifically of the implementation of – mainly technological – innovations through the interplay of socio-technical interactions at different organizational levels[4]. Its description of transition processes is based on a cascade of events that appear at specific organizational levels, and that have to be articulated and linked from one another for the changes to take place. Thus, it is a "complexification" of a determinist perspective, based on recognition of, on the one hand the diversity of "transition pathways" and, on the other hand of sources of change. This is evidenced by a recent modification of this framework, which suggests "a change in the conceptualization of transition dynamics towards a more distributed, multi-source view of change" (Geels, 2018). Initially, this approach was developed in research on sectors such as energy and transport, and was very seldom used on agriculture (Smith, 2006). Its application to the case of agrifood systems has led to a critical discussion of the linear nature of the theory, and its lack of incorporation of changing bio-physical and socio-natural factors. This led to showing, for example, the impacts of convergence effects and of coordination between several different niches (Bui et al., 2016; Elzen et al., 2017; Rossi et al., 2019). Moreover, one of the limits of the MLP, when applied to agrifood systems, is its intrinsic difficulty to convincingly deal with two key characteristics of agrifood systems, that is, the essential distinctiveness and unpredictability of natural and bio-physical processes (both in the production and consumption spheres, which also makes it difficult to define the closure of the system), and the related contingency and place-based context-dependency.

---

[4] More recently developed, the notion of "deep transition" shows how changes across multiple systems become connected and coordinated thus leading to a common directionality in the long run (Schot and Kanger, 2018).

While transition studies address transitions *ex-post*, transition management approaches have been built on the very idea of managing change and *future* transitions. Their aim is to develop management levers and forms of governance capable of bringing about a change in the state of the system (Kemp et al., 2007; Rotmans et al., 2001). Transition management approaches, which have a strong influence in some policy and consultancy circles, have been criticized for mainly aiming at steering the social acceptability of new technologies (Foxon et al., 2009), thus suggesting an alignment between a techno-centric vision of change levers and mechanisms on the one hand, and a determinist perspective on the change process and the way to manage it on the other one.

Sustainability issues have also been amply tackled by the epistemic communities that emerged around socio-ecological approaches, most notably the "resilience" community (Foxon et al., 2009; Ollivier et al., 2018). Unlike transition studies, whose ontological relationship to change revolves around *human* action (translated into rules, devices, strategies, etc.), socio-ecological approaches primarily consider the reactions of *human-environment* configurations and, above all, the high intrinsic uncertainty associated with unpredictable external factors (such as climate). Both of these approaches develop a systemic approach of change, but they base it on different system closures (respectively socio-technical and socio-ecological systems) that relate to different visions of change and interpretations of change mechanisms.

The socio-ecological approaches address change on the basis of a conceptualization of adaptation processes for complex socio-ecological systems in the face of external disturbances. It is the behaviour of the system itself that is studied with regards to its ability to maintain its own structure-function relationship or to transform itself into another socio-ecological system. In the various frameworks proposed in relation to the concept of resilience (Adaptive cycle, Panarchy model, etc.), it is not so much the process of change that is modelled or the nature of the states to be achieved, but rather the adaptive or transformative properties of the system facing external disturbances that drive these changes.

In terms of change management, the resilience community has produced theoretical reflections on management modalities that aim at dealing with the uncertainty inherent to the complexity of socio-ecological systems. What is known as the "adaptive" management mode has thus been formalized as a pragmatic framework for action aimed at acting without being able to anticipate the consequences. The idea is to implement a cycle of action adjustments according to the system's responses (Williams, 2011), in contexts of strong uncertainties and complexity. In contrast to transition management and its focus on the ability to steer long-term changes in the system's functions, adaptive management emphasizes the maintenance of these functions in the face of external change (Foxon et al., 2009). This adaptive management framework has spawned many variations. These variations deal with the scope of adjustment (adapting only actions, reviewing initial targets, or redesigning them with stakeholders) and with the role of stakeholders and their knowledge (from implementation on the basis of ex-ante scientific modelling to a learning process that puts the experiences and skills of stakeholders first). In these adaptive management approaches, the different kinds and scopes of adjustments may thus be anchored in determinist or open-ended perspectives, despite a frequent claim for "open-endness." Considerable ambiguity has thus arisen around the use of the term "adaptive management" which is directly – and often wrongly – associated with an open-ended or at least non-determinist perspective. This adaptive approach has often been adopted and adapted for management issues in agroecology both at productive systems' level (Sabatier et al., 2015) and regarding collective management and governance issues, with diverse conceptualizations of adjustments that were rooted in one or the other of these two perspectives (open-ended and determinist), yet *without it being explicitly stated*. Besides adaptive management, which up to date has been the most widely adopted framework, an emergent literature on reflexive and multi-scalar governance options of socio-ecological systems (including reflexive management, polycentric governance, or global-experimentalist governance, among others) also claim being rooted on more open-ended perspectives to transition management[5].

Even though few authors have explicitly discussed the relation between determinist and open-ended perspectives, recent critical debates over deterministic approaches to transition (as developed in transition studies and transition management) suggest useful arguments for such

---

[5] Global experimentalist governance can also play an important role in agroecological transitions. It is an institutionalized process of participatory and multilevel collective problem-solving, in which the problems (and the means of addressing them) are framed in an open-ended way, and subjected to periodic revision by peers in the light of locally generated knowledge (De Búrca et al., 2014). This favours learning, participation and cooperation (Armeni, 2015). This form of governance can establish processes that enable unimagined alternatives.

a critical engagement. This criticism focuses on the command-and-control perspective, seen as an intrinsic feature of these approaches (Stirling, 2014). Within the notion of transition, Stirling argues, the goal is predetermined by the decisions of a few actors who are generally in a position of power (i.e. dominant actors in the 'regime'). This criticism relies on the fact that the transition process, locked-in and with limited solutions (most often technological in nature) primarily serves the interests of a limited number of actors (as in the energy transition). Stirling thus considers that the notion of transformation is better suited to tackle the type of radical and profound changes that must be implemented to achieve sustainability. This notion of transformation represents a more open and contingent approach, based on the participation of all stakeholders and the principles of deliberative democracy. Hence, the notion of social transformations may come as a more open-ended ontological solution to the more deterministic notion of transitions management. Stirling then draws a dichotomy between the notion of "caring politics of transformation" and that of "controlling management of transition" (Stirling, 2019). Whilst this might be regarded as an "easy" criticism of transition management, to make it highlights the problems of "hard" or "soft" determinism which lies at the central (ex-ante) ontological basis of these approaches.

The argument that on the basis of social emancipation may emerge *more open and pervasive transformations*, rejoins other authors that claim for an emancipatory perspective to the future of food systems (Lacey, 2015) or put forth the notion of localized and/or collective experimentation to support an open and place-based perspective on transitions (Levidow, 2015; Moragues-Faus and Marsden, 2017)[6]. This is likewise the position chosen in approaches that are grounded on the notion of "extended expert community," which consists in including stakeholders in the definition of what to analyse, what to aim for and how to go about it (Funtowicz and Ravetz, 1993; Popa et al., 2015). This position contrasts with "complex system approaches," which claim to take into account the diversity of actors and phenomena in a system while maintaining a reductionist view of the analysis of transition mechanisms and a managerial command and control approach to defining goals in change

---

[6] This is also present in the emerging literature about living labs, with the distinction made by Schliwa and McCormick (2016) between two different strands: user-centric living labs, and citizen-centric living labs.

processes. In the same non-deterministic stance, F. Chateauraynaud and J. Debaz (2017) also develop a "pragmatics" of transformations that takes into account the diversity of possible trajectories, of actors' standpoints, and of interactions between individuals and their "milieux" (living environments).

This short synthesis of the main frameworks in use in sustainable transition approaches shows that each of them is grounded in more determinist or open-ended ontological perspectives on change (even if this is rarely made explicit and can evolve according to processes of adaptation and revision of these frameworks). These two perspectives are therefore at play in agroecological transitions through the very conceptualizations of transitions and transformation process, albeit implicitly since the literature most often contrasts *visions* of agroecology rather than *perspectives of change*. Our hypothesis is thus that these two perspectives are active in transition processes for different objects, at different levels and according to various modalities, and that the analysis of their interactions could be relevant in the understanding of transition mechanisms. Three questions emerge from this hypothesis: How are these two perspectives present and active together in transition and transformation processes? At which scales and levels? Is it rather a matter of tension or of complementarity?

4. The collective process: An attempt to enlighten agroecological transition mechanisms by clarifying our vision on change

With this book collection, we intend to show that both perspectives on change are effectively at play in agroecological transitions, that they interfere jointly in transition processes in complex and potentially unexpected ways, and finally that a close analysis of their role could open new theoretical and operational issues. As said above, our stance is not normative regarding the respective relevance of determinist and open-ended perspectives for AETs. Following Stirling's (2019) arguments, one could argue that more open-ended approaches are needed for dealing with AET, due both to their inherent socio-natural contingency (and thus high level of complexity and uncertainties) and to the fact that they essentially should embody an empowering process, defined as a "politics of care." Agroecology is not a bounded or narrow "technology" as such, as it cannot be consistently replicated over time and space (or place) in a generic, standardized, and linear fashion. Adopting a determinist perspective on agroecology thus renders it weaker and more marginal than it is indeed capable of being. However, we argue that there is not one good open-ended vision of AET and that the reality is instead fuzzy and complex.

We have conceived this book project, following other recent collective books (Elzen et al., 2017), as an exploratory exercise for authors invested in research on agroecological transitions and who were all questioning change in their own way without (for most of them) explicitly addressing the visions of change. The general idea for the book building process was to encourage the authors to take up the topic of the ontological relationships to change in order to "revisit" their research on AET through this critical ontological lens. To initiate and facilitate this process, we organized a 3-days workshop gathering potential co-authors of this book so as to share the idea of the book as a project, to discuss this framework based on determinist and open-ended perspectives, and to work on the first key arguments of the possible chapters for this book. The participants represented a diversity of disciplines (biological, agronomical, and social sciences) working on diverse contexts and situations of transitions, and facing different stakes, objects, and scales.

The use of this framework was suggested as a mean to get away from the normative categorizations associated with different visions of agroecology. It was suggested as an analytical key for examining more precisely how these two "ideal-typical" perspectives describing relationships to change operate, *in combination or in tension*, in the design and implementation of transition projects and in their analysis. In order to launch the collective reflections, a number of open questions were formulated: *Why and how can the two perspectives follow each other, be combined or simply create tensions in situations of agroecological transition? Can we categorize different ways of arranging these perspectives according to transitional situations? Are these arrangements transitional or do they create pathways for transition?* In theory, it is possible to imagine many potential combinations if both perspectives are considered. However, these tensions are highly contingent on transitional situations – that is, contexts, actors and trajectories that will create specific interplay or tensions. From this workshop, a first panel of proposals emerged, as the result of spontaneous sub-group brainstorming. Other proposals then emanated from other colleagues also involved in AETs studies. The chapters presented here are not only the result of a process we initiated and framed, but mostly of an appropriation and exploration work which continued between the co-authors after the workshop. In this way, we wished to engage a creative process which, as a participant in our initial workshop put it in the conclusion session, is itself a combination of a determinist perspective (through our guidance to work on specific objectives based on our framework) and an open-ended one (through the facilitation process which allowed authors to draw and develop unexpected themes and co-authorships).

The issues and objects tackled along the 11 chapters range from agricultural experimentation to varietal innovation and agroecological policies, and from the scale of an individual farm to that of territorial food systems, as well as comprising different national contexts. Together, the chapters reflect the diversity we expected when starting the collective process, both in the way visions on change are tackled when studying agroecological transitions, and in the extent to which the different authors adopt the suggested framework.

In a first group of chapters, the authors have used this determinist versus open-ended perspective framework as an analytical grid for a self-critical "re-reading" of their research issues and objects, with a varying degree of reinterpretation of the two perspectives.

Mireille Navarrete, Hélène Brives, Maxime Catalogna, Amélie Lefèvre and Sylvaine Simon analyse the articulation of the two perspectives around design processes. They deconstruct the idea of a systematic alignment between research-driven design processes and determinism and a farmers' driven one and open-endedness, but support that determinist phases are inevitable and necessary in any design process. They identify, based on different cases of experimentations for cropping systems, three patterns of intertwining the two perspectives: (i) a strict succession of the two, (ii) a progressive replacement of one perspective by the other one or, (iii) the development over time of the two perspectives in parallel. These patterns appear as respective responses to uncertainties and knowledge gaps, to disparity between actors or to the way actors articulate the short and long terms in the way they set up their experimentations.

Sophie Tabouret, Claire Lamine and François Hochereau analyse the evolution of varietal breeding in perennial plants and show the contrast between a determinist perspective translated by the notion of ideotype, and a more open-ended one. This characterizes a "paradigm shift" and translates into participatory approaches that are based on the establishment of a larger socio-technical network. However, they show that this general overview is embedded in a more complex process at the project-level in this particular case of perennial crops, where determinist perspectives are also at play in diverse combinations with open-ended ones, with different framings of the socio-technical network involved.

Marianne Hubeau, Martina Tuscano, Fabienne Barataud and Patrizia Pugliese put at test the determinist versus open-ended perspective framework in the analysis of trajectories of territorial agri-food projects. Studying how the objectives of four multi-actor, multilevel and multi-scale agri-food projects (in Belgium, France and Italy) are defined, how these projects are governed and how processes of consideration and/or exclusion of actors are managed, they show how these two perspectives highlight the *evolutive* feature of the multi-actors' process at play, giving consistency to the context-dependency of these transitions.

Claire Lamine, Claudia Schmitt, Juliano Palm, Floriane Derbez, and Paulo Petersen analyse the way innovative public policies' instruments in France and Brazil may support *a priori* open-ended agroecological transitions at the scale of farmers' groups in contrast to more conventional instruments. Here too, the analysis shows the entanglement of the two perspectives. Indeed, the authors show how in their content, these instruments combine determinist and open-ended perspectives to agroecological transitions through the frame they provide to the groups' projects – the groups may *define freely* their objectives and their own indicators of transition, but they are *obliged* to establish them and monitor them throughout their projects. These instruments thus combine expected normative effects with unexpected performative ones linked to the way the groups use them and adapt their potentialities to their own situation.

Moacir Darolt, Juliette Anglade, Pascale Moiti-Maïzi, Claire Lamine, Florette Rengard, Vanessa Iceri, Amélie Genay, and Cristian Celis also explore the tensions and/or articulation between the two perspectives regarding teaching and training agroecology. They show – comparing professional teaching programs and programs implemented by grass-root movements in « metropolitan » France and Brazil – that pedagogies range from diffusionist stances (determinist) to socio-constructivist ones (open ended). They suggest, however, that synergic arrangements could be explored between determinist and open-ended perspectives on how to relate knowledge and action (rather than simply opposing scientific and experiential knowledge).

Marc Barbier, Sarah Lumbroso, Jessica Thomas and Sébastien Treyer analyse the links between relationships to change and the activities devoted to the «manufacture of the future», based on six case studies that gather foresight exercises, public policies programs and local collective projects. They link the determinist and open-ended perspectives to two archetypes of transition pathways: the optimization pathway (linked to the current socio-technical regime of agri-food systems) and the transformative pathway (referring to radical and systemic changes as claimed by many agroecological actors and scholars). This leads them to also distinguish two types of commitment to change for an agroecological transition: an "evidence-based" transition, relying on techno-scientific levers and evidence to be mobilized to define the expected changes; and an "experiential/learning-based" transition, relying on a concrete and direct implementation of actions and experimentation in order to build a transition "in itinere."

The second group of chapters of the book does not centrally refer to the determinist and open-ended perspectives but suggest a reflection on the visions of change in their work. Two chapters on livestock system transitions show how change is not always oriented and prescriptive (as in design processes for example), but also largely constrained and contingent. In that way, they both highlight the tension between adaptation or transition or, more largely, the permeability and linkages between the two processes.

Vincent Thénard, Gilles Martel, Jean-Philippe Choisis, Timothée Petit, Sébastien Couvreur, Olivia Fontaine, Marc Moraine interpret transitions as the way that a productive system will evolve within a territorial space of constraints and opportunities, thus addressing transition as an « adaptation ». They develop a conceptual framework to describe livestock systems' transitions through their ability to access and combine a large range of resources (ecological, technical, and socio-economical) of their territories. They apply this framework to different case studies of livestock systems in metropolitan France and in the Reunion island, showing the different types of recombination of resources that the systems operate as the result of the crossing of actors' strategies and territorial contexts.

Charles Henri Moulin, Laura Etienne, Magalie Jouven, Jacques Lasseur, Martine Napoléone, Marie-Odile Nozières-Petit, Eric Vall and Arielle Vidal formalize a specific change issue for agropastoral systems as they need to transition in the face of territorial dynamic pressure, while maintaining themselves as agroecological systems. Using French and African case studies, they show how transitions operate through arrangements of resources and actors' interactions at different levels, highlighting the relevance of a multi-scalar analysis (at the scales of the livestock system, farm, household, territory, local food system) to build an accurate assessment of transition pathways within agroecology. The adaptation is discussed here as the ability of the system to remain within agroecology.

Based on a transversal analysis of four cases studies of research-action dealing with agroecological transitions, the chapter by Pierre Stassart, Antoinette Dumont, Corentin Hecquet, Stéphanie Klaedtke, Camille Lacombe and Matthieu de Nanteuil highlights the role of normative dimensions at stake in change processes. The authors explore how transition processes perform through the collective building of what these authors call a "shared normative model," resulting from the resolution of tensions and ambiguity between actors, who all share a quest for justice. Drawing on the specificities of the different "situations," they identify three normative frames defined by the way they envision justice. They relate these to the three ethics of compromise, of capabilities, and of recognition, and clearly argue for an open-ended perspective to transition, considered as induced from the choice of a transdisciplinary stance.

The chapter by Divya Sharma and Barbara Van Dyck also focusses on visions at play on change, viewed as an object of struggle for social movements for food sovereignty. They resituate the coexistence of visions within the place-based histories of agroecology movements, through a conversation between the two authors about the trajectories of two such social movements in the north Indian state of Punjab and in Belgium. For the authors, thinking through the lens of the other – trans-local movement conversations – may open agroecology movements in a way (that may be qualified as open-ended) that does not contribute to relegating them to marginalized groups, but instead create space for building solidarity.

Lastly, Michael Bell and Stéphane Bellon propose an analysis of the transformative capacity of agroecology based on the analysis of rhetorical strategies. Through the concept of boundary strategies, the authors depict how strategies are context-dependent and defined in relation to other narratives (internal or external, against or with). They show how borders are defined by these narratives and analyse the strength (weak/strong) and permeability (close/open) of these boundaries. The way that these boundaries evolve through narratives define trajectories that the authors describe as determinist when boundaries keep close and strong and open-ended when boundaries are relatively open and weak.

The diversity of issues presented here shows the importance of taking visions of change as a key dimension in sustainable agroecological transitions. The framework that has been suggested, refined, and discussed along the collective process of construction of this book, over nearly two years, has proved useful to clarifying the relationships to change that play out in agroecological transitions. Far from resulting in a clear-cut dichotomic characterization, this collective process and the diverse case studies and analyses carried out here lead to identifying a diversity of possible articulations and combinations of determinist and open-ended perspectives. Along the chapters, it has become clear that these two perspectives can appear one after the other, in combination, or in tension, and that these tensions and articulations are instrumental to any real "progress" along agroecological transition pathways, whose diversity has to be kept open. We also believe that beyond its heuristic utility for understanding the perspectives of the actors involved in the agroecological transition and/or its analysis, the framework we have explored in this book could have a "reflexive" role individually or collectively in action research situations.

References

Altieri, M. A., Nicholls, C. I., Henao, A., Lana, M., 2015. Agroecology and the Design of Climate Change-Resilient Farming Systems, *Agronomy for Sustainable Development*, 35, 3, 869–90. https://doi.org/10.1007/s13593-015-0285-2.

Armeni, C., 2015. Global Experimentalist Governance, International Law and Climate Technologie, *International and Comparative Law Quarterly*, 64, 875–904.

Berthet, E., Barnaud, C., Girard, N., Labatut, J., Martin, G., 2016. How to Foster Agroecological Innovations? A Comparison of Participatory Design Methods, *Journal of Environmental Planning and Management*, 59, 2, 280–301. https://doi.org/10.1080/09640568.2015.1009627.

Bui, S., Cardona, A., Lamine, C., Cerf, M., 2016. Sustainability Transitions: Insights on Processes of Niche-Regime Interaction and Regime Reconfiguration in Agri-Food Systems, *Journal of Rural Studies*, 48, 92–103. https://doi.org/10.1016/j.jrurstud.2016.10.003.

Chateauraynaud, F., Debaz, J., 2017. *Aux bords de l'irréversible. Sociologie pragmatique des transformations*, Paris, Petra.

Cowan, R., Gunby, P., 1996. Sprayed to Death : Path Dependence, Lock-in and Pest Control, *Economic Journal*, 106, 436, 521–43.

Dalgaard, T., Hutchings, N. J., Porter, J. R, 2003. Agroecology, Scaling and Interdisciplinarity, *Agriculture, Ecosystems & Environment*, 100, 1, 39–51. https://doi.org/10.1016/S0167-8809(03)00152-X.

De Búrca, G., Keohane, R. O., Sabel, C., 2014. Global Experimentalist Governance, *British Journal of Political Science*, 44, 477–486.

de Molina, M. G., 2013. Agroecology and Politics. How to Get Sustainability? About the Necessity for a Political Agroecology, *Agroecology and Sustainable Food Systems*, 37, 45–59. https://doi.org/10.1080/10440046.2012.705810.

Dosi, G., 1982. Technological Paradigms and Technological Trajectories, *Research Policy*, 11, 147–62.

Duru, M., Therond, O., Fares, M., 2015. Designing Agroecological Transitions; A Review, *Agronomy for Sustainable Development*, 35, 4, 1237–57. https://doi.org/10.1007/s13593-015-0318-x.

Elzen, B., Augustyn, A. M., Barbier, M., van Mierlo, B., 2017. AgroEcological Transitions: Changes and Breakthroughs in the Making. https://doi.org/10.18174/407609.

Foxon, T. J., Reed, M. S., Stringer, L. C., 2009. Governing Long-Term Social–Ecological Change: What Can the Adaptive Management and Transition Management Approaches Learn from Each Other? *Environmental Policy and Governance*, 19, 1, 3–20. https://doi.org/10.1002/eet.496.

Francis, C., King, J., Lieblein, G. Breland, T. A., Salomonsson, L., Sriskandarajah, N., Porter, P., Wiedenhoeft, M., 2009. Open-Ended Cases in Agroecology: Farming and Food Systems in the Nordic Region and the US Midwest, *The Journal of Agricultural Education and Extension*, 15, 4, 385–400. https://doi.org/10.1080/13892240903309645.

Funtowicz, S., Ravetz, R., 1993. Science for the Post-Normal Age, *Futures*, 7, 25, 739–755.

Geels, F. W., 2002. Technological Transitions as Evolutionary Reconfiguration Processes: A Multi-Level Perspective and a Case-Study, *Research Policy*, 3, 8–9, 1257–1274. https://doi.org/10.1016/S0048-7333(02)00062-8.

Geels, F. W., 2018. Low-Carbon Transition via System Reconfiguration? A Socio-Technical Whole System Analysis of Passenger Mobility in Great Britain (1990–2016), *Energy Research & Social Science*, 46, 86–102. https://doi.org/10.1016/j.erss.2018.07.008.

Geels, F. W., Schot, J., 2007. Typology of Sociotechnical Transition Pathways, *Research Policy*, 36, 3, 399–417. https://doi.org/10.1016/j.respol.2007.01.003.

Giraldo, O. F., Rosset, P. M., 2018. Agroecology as a Territory in Dispute: Between Institutionality and Social Movements, *The Journal of Peasant Studies*, 45, 3, 545–64. https://doi.org/10.1080/03066150.2017.1353496.

Girard, N., Magda, D., 2018. The Interplays Between Singularity and Genericity of Agroecological Knowledge in a Network of Livestock Farmers, *Revue d'anthropologie des connaissances*, 12, 2, 2, 199–228.

Jullien, F., 2004. *A Treatise on Efficacy: Between Western and Chinese Thinking*, translated by Janet Lloyd, Honolulu, University of Hawaii Press.

Kemp, R., Loorbach, D., Rotmans, J., 2007. Transition Management as a Model for Managing Processes of Co-Evolution towards Sustainable Development, *International Journal of Sustainable Development & World Ecology*, 14, 1, 78–91. https://doi.org/10.1080/13504500709469709.

Lacey, H., 2015. Food and Agricultural Systems for the Future: Science, Emancipation and Human Flourishing, *Journal of Critical Realism*, 14, 3, 272–86. https://doi.org/10.1179/1572513815Y.0000000002.

Lamine, C., 2017. *La Fabrique Sociale de L'écologisation de L'agriculture*, Marseille, La Discussion.

Levidow, L., 2015. European Transitions towards a Corporate-Environmental Food Regime: Agroecological Incorporation or Contestation? *Journal of Rural Studies*, 40, 76–89. https://doi.org/10.1016/j.jrurstud.2015.06.001.

Liebowitz, S., Margolis, S. E., 1995. Policy and Path Dependence: From QWERTY to Windows 95, *Regulation*, 18, 33.

Loorbach, D., Frantzeskaki, N., Thissen, W., 2011. A Transition Research Perspective on Governance for Sustainability, in Jaeger, C., Tàbara, J., Jaeger, J. (Eds.), *European Research on Sustainable Development*, Berlin, Heidelberg, Springer. https://doi.org/10.1007/978-3-642-19202-9_7.

López-i-Gelats, F., Di Masso, M., Binimelis, R., Rivera-Ferre, M. G., 2016. Agroecology, in *Encyclopedia of Food and Agricultural Ethics*, 1–6, Netherlands, Springer.

Marsden, T. K., 2017. *Agri-Food and Rural Development: Sustainable Place-Making*, London, Bloomsbury.

Meek, D., 2016. The Cultural Politics of the Agroecological Transition, *Agriculture and Human Values*, 33, 2, 275–90. https://doi.org/10.1007/s10460-015-9605-z.

Méndez, V. E., Bacon, C. M., Cohen, R., 2013. Agroecology as a Transdisciplinary, Participatory, and Action-Oriented Approach, *Agroecology and Sustainable Food Systems*, 37, 1, 3–18. https://doi.org/10.1080/10440046.2012.736926.

Meynard, J. M., Jeuffroy, M. H., Le Bail, M., Lefèvre, A., Magrini, M. B., Michon, C., 2016. Designing Coupled Innovations for the Sustainability Transition of Agrifood Systems, *Agricultural Systems*, August. https://doi.org/10.1016/j.agsy.2016.08.002.

Moragues-Faus, A., Marsden, T., 2017. The Political Ecology of Food: Carving 'Spaces of Possibility' in a New Research Agenda, *Journal of Rural Studies*, 55, October, 275–88. https://doi.org/10.1016/j.jrurstud.2017.08.016.

Norder, L. A., Lamine, C., Bellon, S., Brandenburg, A., Norder, L. A., 2016. Agroecology, Polysemy, Pluralism and Controversies, *Ambiente Sociedade*, 19, 3, 1–20. https://doi.org/10.1590/1809-4422ASOC129711V1932016.

Ollivier, G., Magda, D., Mazé, A., Plumecocq, G., Lamine, C., 2018. Agroecological Transitions: What Can Sustainability Transition Frameworks Teach Us? An Ontological and Empirical Analysis, *Ecology and Society*, 23, 2. https://doi.org/10.5751/ES-09952-230205.

Popa, F., Guillermin, M., Dedeurwaerdere, T., 2015. A Pragmatist Approach to Transdisciplinarity in Sustainability Research: From Complex Systems Theory to Reflexive Science, *Futures*, Advances in Transdisciplinarity, 2004–2014, 65, January, 45–56. https://doi.org/10.1016/j.futures.2014.02.002.

Possas, M. L., Salles-Filho, S., da Silveira, J. M., 1996. An Evolutionary Approach to Technological Innovation in Agriculture: Some Preliminary Remarks, *Research Policy*, 25, 933–45.

Purdy, J., 2015. *After Nature: A Politics of the Anthropocene*, Cambridge MA, USA, Harvard University Press.

Purdy, J., 2019. *This Land in Our Land: The Struggle for a New Commonwealth*, Princeton, USA, Princeton University Press.

Rip, A., Kemp, R., 1998. Technological Change, in Rayner, S., Malone, E. L. (Eds.), *Human Choice and Climate Change. Vol. II, Resources and Technology*, Columbus, OH, Battelle Press, 327–399. http://doc.utwente.nl/34706/.

Rivera-Ferre, M. G., 2018. The Resignification Process of Agroecology: Competing Narratives from Governments, Civil Society and Intergovernmental Organizations, *Agroecology and Sustainable Food Systems*, 42, 6, 666–85. https://doi.org/10.1080/21683565.2018.1437498.

Rossi, A., Bui, S., Marsden, T., 2019. Redefining Power Relations in Agrifood Systems, *Journal of Rural Studies*. https://doi.org/10.1016/j.jrurstud.2019.01.002.

Rotmans, J., Kemp, R., van Asselt, M., 2001. More Evolution than Revolution: Transition Management in Public Policy, *Foresight*, 3, 1, 15–31. https://doi.org/10.1108/14636680110803003.

Sabatier, R., Oates, L. G., Jackson, R. D., 2015. Management Flexibility of a Grassland Agroecosystem: A Modeling Approach based on Viability Theory, *Agricultural Systems*, 139, 76–81. https://doi.org/10.1016/j.agsy.2015.06.008.

Schliwa, G., McCormick, K., 2016. Living Labs: Users, Citizens and Transitions, in *the Experimental City*, London, UK, Routledge, 163–178.

Schot, J., Kanger, L., 2018. Deep Transitions: Emergence, Acceleration, Stabilization and Directionality, *Research Policy*, 47, 6, 1045–59. https://doi.org/10.1016/j.respol.2018.03.009.

Smith, A., 2006. Green Niches in Sustainable Development: The Case of Organic Food in the United Kingdom, *Environment and Planning C: Government and Policy*, 24, 3, 439–58. https://doi.org/10.1068/c0514j.

Spaargaren, G., Oosterveer, P., Loeber, A., 2012. *Food Practices in Transition: Changing Food Consumption, Retail and Production in the Age of Reflexive Modernity*, New York, Routledge.

Stirling, A., 2014. *Emancipating Transformations: From Controlling 'The Transition' to Culturing Plural Radical Progress*, STEPS Working Paper 64, Brighton, STEPS Centre.

Stirling, A., 2019. Sustainability and the Politics of Transformations: From Control to Care in Moving beyond Modernity, *What Next for Sustainable Development?* July. https://www.elgaronline.com/view/edcoll/9781788975193/9781788975193.00023.xml.

Vanloqueren, G., Baret, Ph. V., 2008. Why Are Ecological, Low-Input, Multi-Resistant Wheat Cultivars Slow to Develop Commercially?

A Belgian Agricultural 'lock-In' Case Study, *Ecological Economics*, 66, 2–3, 436–46. https://doi.org/10.1016/j.ecolecon.2007.10.007.

Williams, B. K., 2011. Adaptive Management of Natural Resources – Framework and Issues, *Journal of Environmental Management*, Adaptive Management for Natural Resources, 92, 5, 1346–53. https://doi.org/10.1016/j.jenvman.2010.10.041.

Wilson, C., Tisdell, C., 2001. Why Farmers Continue to Use Pesticides Despite Environmental, Health and Sustainability Costs, *Ecological Economics*, 39, 449–62.

Intertwining deterministic and open-ended perspectives in the experimentation of agroecological production systems: A challenge for agronomy researchers

MIREILLE NAVARRETE, HÉLÈNE BRIVES, MAXIME CATALOGNA, AMÉLIE LEFÈVRE, SYLVAINE SIMON

1. Introduction

For decades, change in agriculture has stemmed primarily from the development of technical innovations such as new cultivars, machinery, and synthetic inputs. During this "modernization" phase, researchers and technical advisors assumed that farmers would adopt new techniques and knowledge that they had developed and disseminated. This top-down linear process was criticized extensively (Klerkx and Leeuwis, 2008; Duru et al., 2015) and was gradually enriched through the involvement of farmers in the innovation process (Salembier et al., 2018). With expanding interest in agroecology, the ways in which knowledge and innovations are generated are now receiving close attention, as recognition increases with respect to the numerous and complex interactions between the components of farming systems, between groups of living organisms, and between short- and long-term dynamics. Our base of scientific knowledge remains insufficient to understand the consequences of these complex interactions and dynamics on agroecological systems and to develop predictive models to help farmers manage their farming systems in an agroecological way. In particular, the intensity and speed at which natural regulations may occur are not fully predictable. For example, will the sowing of a particular pest-trap crop be sufficient to control the pest all along its development and provide a satisfying production? Moreover, agroecological knowledge is context specific, that is, the performance of a particular practice heavily depends on the local ecological conditions in which it is applied. Grounding farming system approaches on ecosystem services that are supported by ecological processes and biodiversity requires profound redesign rather than simple adaptation. Duru et al. (2015) refer to this alternative path forward as the "strong ecological modernization of agriculture," which focuses more attention to farmers' learning and on-farm innovation processes in order to overcome problems related to uncertain and situated ecological processes (Doré et al., 2011; Prost et al., 2016).

On-farm experimentation, where farmers create a specific situation to be observed in their own farming context, is an important way to generate practical knowledge and favor technical change (Leitgeb et al., 2014). On-farm experimentation is also a way to address the uncertainty and complexity associated with biodiversity-based agriculture (Duru et al., 2015). In science, experimentation has been a foundational way to produce knowledge, especially in agronomy where it has enabled the identification of generic laws regarding field and crop functions, and hence the optimization of techniques and the prediction of their effects (Maat and Glover, 2007). These authors note that "experimentation plays a crucial role in connecting the academic discipline of agronomy with agricultural practice" (p.132). As agronomic topics and methods are questioned by the agroecological transition, re-opening the debate on best practices for impactful experimentation is a timely discussion. In addition to the predefined classical agronomic experiments which were designed to produce generic knowledge under controlled conditions, alternative experimental approaches have emerged, which occur in an open framework to support both researchers' and farmers' learning processes (Cardona et al., 2018). In this chapter, we analyzed several experimentation processes of agroecological systems, through the lens of deterministic and open-ended perspectives. It gave a fresh look at our role as researchers in agronomy involved in experimentation and engaged us *a posteriori* in a self-reflection process, with the help of a sociologist. First, based on a literature analysis, we delve deeper into the exploration of how experimentation is questioned by agroecology and draw two broad conceptions of deterministic and open-ended perspectives in experimentation (Section 2). We then present the empirical material, coming from four French case studies on experimentation falling within both deterministic and open-ended perspectives (Section 3). We describe different ways by which the two perspectives complement each other (Section 4), and we finally discuss the interests and limits of such a combination (Section 5).

2. Reconsidering experimentation on farming systems with agroecology

For a long time, agronomic experimentation[1] only consisted in testing hypotheses based on the current scientific knowledge and established causal links between actions and effects. The experimental layouts consisted in creating controlled environments (in laboratories or fields) to compare experimental treatments (Maat and Glover, 2007), which were then analyzed with a highly formalized protocol. For example, in factorial experiments, which are emblematic of that period, the biophysical environment is split into a limited number of factors, highly controlled to avoid potential interactions with factors considered as minor, and unstudied. Almost all decisions are set prior to the implementation of the experiment to fit to statistical requirements (location of the treatments, experimental protocols). The behavior of the experimental system is assumed to be representative of real farming situations, so that recommendations usable by farmers in a large range of conditions can be defined. Outside of this monolithic definition of acceptable experimentation, new experimental approaches progressively emerged with the need to address systemic issues, particularly in France at the end of the 20th century. These new approaches realized on-farm or on-station involve stakeholders in the experimental decisions and provide opportunities for knowledge-generation through improvement feedback loops rather than predetermined protocols. Two types of experimentation deserve special attention because they offer strong potential for reconciling agronomic experimentation with the challenges of agroecology: system experiments and farmers' experiments. **A system experiment** consists of designing and implementing what is hypothesized to be the optimal set of crop sequence and technical management to reach certain predetermined goals (e.g., agronomic, environmental and economic goals), in order to assess their performance within a given context (Debaeke et al., 2009; Meynard et al., 2012). Such experimented systems are not necessarily fixed throughout time, but can be adapted over years according to biotic, abiotic and social contexts, to the development of new knowledge or techniques, or to day-to-day management (Lechenet et al., 2017). System

---

[1] In this chapter, the term "experimentation" refers to the whole process of experimenting including its social and financial dimensions, whereas "experiment" refers to the practical layout.

experiments are thus a major breakthrough in approach as compared to the classical factorial perspective where framework and modalities are predetermined and a complex system is broken down into smaller units to be experimented. The second type is the **experiments carried out by farmers** to identify efficient cropping practices to implement on their own farms (Johnson, 1972; Saad, 2002). While farmers' experiments have existed from the onset of agriculture, renewed interest has emerged with the perspective of agroecology, as it provides farmers with tools to adapt their systems to uncertainty and to build situated knowledge (Navarrete et al., 2018).

From the perspective of agroecology, optimizing ecological processes in crop management is complicated by the lack of scientific knowledge, the close dependence of such processes on site-specific environment, and the numerous interactions of biological, chemical, and physical factors with cropping practices. All these elements result in a high level of variability on dynamics overall, which impedes any reliable prediction. We therefore consider as irrelevant the application of rigid experimental protocols only. A more relevant approach would be to define general objectives as support for decision making, and to modify decisions along the way based on on-going observations of the system. Additionally, to increase the relevancy and utility of data generated from experiments, we advocate that such day-to-day management of the experimented cropping systems should be considered as a rich contribution to experimenters' learning. And finally, we consider that experimentation could broaden learning through social interaction between scientific and non-scientific actors, professional and non-professional experimenters.

In line with the main theme of the book, we analyze agroecological experimentation through the lens of deterministic and open-ended perspectives. Agronomic experimentation carries in itself a vision of a technical change that can be steered by humans, from an existing biophysical system (often considered as a reference) to a new and improved one which is expected to better reach the specified aims. Experimentation thus consists in three main activities: imagining the new system, implementing it practically, and observing its properties to check to which degree it satisfies the aims. With agroecology, we pay specific attention to the question of predictability of the systems to experiment and the degree to which experimental decisions can be planned in advance. As an initial approach, we consider that the deterministic perspective refers to experiments where most elements (goals, objectives, type of knowledge to build, hypotheses to test, assessment criteria, data collected, etc.) are fully stabilized from the outset. By contrast, an open-ended perspective requires an iterative approach where both the goals and the means to reach them are intentionally adapted based on system observations and social exchanges. The overall strategic implementation of practices is gradually refined on the basis of ecological and social dynamics even if different from the practices thought out before the experiment. One might imagine that the experiments undertaken by researchers would belong essentially to the deterministic perspective, whereas farmers, who are not expected to scientifically validate their results to the same degree, would adopt a more open-ended perspective for experimentation. In this chapter, we show that experiments on agroecological systems largely transcend this rough categorization. **More precisely, we assume that deterministic and open-ended perspectives coexist in experimentation and that such coexistence is linked to the specificities of agroecological systems.** Therefore, such a proposal carries along with it the need for a paradigm shift in knowledge production for agronomy.

3. Methods

We **cross-analyzed four case studies (CS) of experimentations located in France** (Tab. 1). The four experimentations have been developed with a participatory approach of agroecological transition, valuing singular and local knowledge of farmers coming from practical experience (Berthet et al., 2016). In all the CS, experimentation was the place for a specific dialogue organized between farmers and researchers.

The first two CS describe **multi-annual system experiments implemented at research stations and dedicated to the design of agroecological farming systems in a step-by-step process**. The two cases enabled the analysis of why and how an open-ended perspective was embedded in an initially deterministic one, and which specific elements of the experiments were impacted. Step-by-step design system experimentation (Coquil et al., 2014) emerged recently as a new twist of system experimentation, where researchers, technical advisors and farmers share their knowledge to manage and analyze the cropping system, and progressively adapt it to fit uncertainty (e.g., unacceptable development of a pest, lack of efficiency of a technique, etc.), in a learning-by-doing approach (Meynard et al., 2012; Navarrete et al., 2017). In the two CS, the experiments were carried out at INRAE experimental stations. The first CS, conducted at Gotheron station, dealt with apple orchards using two different approaches: (a) the BioREco, a pioneering system experiment on fruit, assessed 9 planned cropping strategies combining 3 types of crop management (conventional, low-input, organic) and 3 apple cultivars differing in disease susceptibility (Simon et al., 2011; Alaphilippe et al., 2013); and (b) the agroecological Zero pesticide orchard project (in short, the Z project) aimed at redesigning and assessing a pest suppressive fruit production area by strongly reinforcing ecosystem services. The second CS, 4SYSLEG, at the experimental station of Alénya, focused on the evaluation of 4 vegetable cropping systems designed to avoid or greatly reduce synthetic pesticides while meeting the standards of a specific food value chain (organic production or conventional, long supply chain or local direct sale) (Lefèvre et al., 2015; Perrin and Lefèvre, 2019).

These first two system experiments, which had been conducted by researchers even if farmers were involved, were compared to **two additional CS where the experiments were conducted by or with farmers.** The goal was to determine if the strong involvement of non-professional experimenters led to a more open-ended way of experimenting. In the CS on farmers' experimentation in the Drôme *département*, we analyzed the experiments initiated by 17 individual farmers over the past ten years. They dealt with no-till, cover-cropping or conservation agriculture on arable crops, and the enhancement of natural enemies of pests or biocontrol practices on vegetable crops. The short and long-term experimentation processes previously characterized (Catalogna, 2018; Catalogna et al., 2018) were re-analyzed here according to the deterministic or open-ended perspectives. The last CS consists of on-farm multi-annual system experiments closely associating farmers and researchers on 4 arable farms through the SOIL network in the Isère *département*. The experiments had two aims: to test and assess some indicators on the biological status of soils that had previously been developed by the researchers, and to describe how soil health evolved over the years with conservation agriculture practices (Boidron, 2018).

The present analysis was based on a reflexive and retrospective review of how the experimentations on the 4 CS were carried out. We looked at how the experimental layouts and their on-going management were decided, implemented, and adapted in the course of the projects, and what the reasons were for stabilizing some decisions or conversely making them evolve. After detecting situations falling within the scope of open-ended and deterministic perspectives, we established whether each perspective was devoted to one specific phase in the experimentation process, to one specific type of stakeholder, and to specific elements of the experimental layout. Then we analyzed the coexistence between the two perspectives. How concretely do open-ended and deterministic perspectives articulate to one another? Do they fall within complementarity or competition? Do they sometimes lead to inconsistent decisions in experiments?

4. A large diversity in the way deterministic and open-ended perspectives coexist in experimentation

4.1. From a deterministic experimentation to a combination of the two perspectives

The system experiment on apple orchards in the BioREco project was initially built by researchers in agronomy and entomology with a deterministic perspective (Fig. 1a), even though farmers and extension agents were involved during some aspects of project development and implementation. The researchers formalized objectives, constraints (on soil, climate, field surface, machinery and labor availability), economic context (sales channel targeted), the experimental layout, agroecological practices and a specific set of decision rules for each experimented cropping system. The initial experimental layout was designed to compare the 9 experimental treatments over time, with a scientific approach even if there were no replicates as in factorial experiments. This type of framework, set from the outset, was very useful for managing the experiments and evaluating the systems. However, 8 years after planting, the management of the conventional systems was changed to include new knowledge learned and exchanges with stakeholders (Fig. 1a). For example, the conventional systems accounted for the farmers' most common practices in the area as basis for systems comparisons. Updates to those systems were implemented in 2013 as some new practices, such as mating disruption, had largely been adopted by the local farmers and became standard practice. Hence, from that year onwards, the experiment became more open-ended to tackle farmers' expectations, include recent innovations and improved practices (e.g., organic fertilization practices were adopted in low-input systems to limit environmental impacts). However, despite such flexibility, the experimented cropping systems based on monoclonal high-density orchards still failed to drastically decrease pesticide

Tab. 1: Short description of the four case studies.

Case study	Step-by-step design system experiments in research station		On-farm experiments	
	From BioREco to Z project (2004–ongoing)	4SYSLEG project (2012–2018)	On-farm experiments by individual farmers (2016–2018)	Joint experiments between researchers and a farmer network in SOIL network (2015–ongoing)
Location	Drôme, INRAE Experimental Station of Gotheron	Pyrénées-Orientales, INRAE Experimental station of Alénya	Drôme	Isère
Socio-economic context	Experiments funded by the French Ecophyto program searching for robust cropping systems using 50% fewer pesticides than historical available references		Farmers interested in developing agroecological practices on their farm	Project funded by Fondation de France
Agroecological issue	Reduce (BioREco) or withdraw (Z project) pesticide use on apple orchard	Reduce or withdraw pesticide use on vegetable sheltered crops	Develop agroecological practices on arable (A) and vegetable (V) crops	Increase soil biological fertility on arable cropping systems
Agroecological practices experimented	BioREco: resistant cultivars, biocontrol, agronomic practices Z: natural and cropped biodiversity, plant spatial arrangement, sanitation	Spatial and temporal crop diversification, natural regulations	A: Conservation agriculture, cover crops V: Biocontrol, reduction of soil tillage	Conservation agriculture, cover crops, crop diversification
Experimental layout	BioREco: 9 cropping systems Z: 1 agroecosystem No replicate Pluriannual data collection	4 cropping systems No replicate Pluriannual data collection	1–19 annual experiments per farmer, between 2003 and 2017 No replicate	4 cropping systems on 4 farms No replicate Pluriannual data collection

use. The following Z Project orchard was consequently completely redesigned. In this multi-actor project, there was no preconception about spatial design or management practices; experimental goals were to control pests through the promotion of ecological processes, without any chemical pesticide. Agronomic and economic performances to target remained open and determined as the project progressed. Conversely, some features of the orchard needed to be formalized before putting the experimental prototype into practice (e.g., orchard shape, cultivar choice). Moreover, researchers felt the need to formalize practical guidelines to steer how to reinforce targeted ecosystem services through practices. They were used not only to manage the experimental layout in a learning-by-doing approach, but also for generic purposes and to share the approach beyond the project partners.

4.2. A planned coexistence between deterministic and open-ended perspectives to tackle uncertainty

In the 4SYSLEG project at INRAE Alénya experimental station, a system experiment was implemented on vegetables. During the 6 years of the project, researchers in agronomy assumed that both strategic planning and tactical adaptation would be alternatively or simultaneously useful, not only to design the four agroecological crop management strategies, but also to implement them in field plots, and to continuously assess and improve them. At the initial stages of project development, a deterministic perspective was adopted to meet the criteria of the Ecophyto call for proposals in 2012 (Fig. 1b), which aimed to specify clear and operational sets of objectives for the crop protection strategies. Thus, quality specifications from marketing chains were translated into a range of priority functions that the cropping systems experimented were expected to fulfil, with the help of invited farmers and extension agents. Experimenters expressed these functions as practical agronomic management at strategic and tactical levels and for each cropping system. On the strategic level, for the low-pesticide cropping system devoted to the direct sale market, it was expected to produce moderate volumes of vegetables but with a wide assortment of vegetables throughout the year. As minor damage to the vegetables is accepted for direct sale, it was decided to continuously enhance natural regulation of pests and diseases, using synthetic pesticides as little as possible and to spend as little time as possible on crop protection. Thus, cropping high levels of spatio-temporal plant diversity, prohibiting deep tillage and soil solarization were identified as key agroecological strategic practices. To bear initially determined sets of objectives in mind for 6 years in each of the four experimental systems, the experimenters summarized each overall strategy into a short slogan (e.g., "Natural balances work with me" for the low-pesticide direct sale system). It framed their decisions and choices, and in particular tactical decisions to sustain natural regulation mechanisms.

As knowledge and innovative solutions for agroecological crop protection were dispersed among many stakeholders, the 4SYSLEG experimenters organized an open-ended approach to fill knowledge gaps with farmers and technical advisers (Fig. 1b). At the onset of the experiment, researchers listed the main damaging pests and diseases feared for each crop or intercrop in order to anticipate practical solutions. They also adopted an adaptive stance for unexpected situations (e.g., new available biocontrol tool, unexpected sanitary or climatic conditions). Thus, even though the main framing of the agroecological crop management strategies was planned in the first stages of the project in a deterministic way, they regularly mobilized farmers and extension agents' expertise during collective workshops to address specific questions, for example, when initially planned objectives could not be met as expected. For example, aligning practices with the slogan "Natural balances work with me" proved difficult for some pest susceptible crops, and a trade-off had to be found between antagonist goals: after a few months, the experimenters removed the requirement to reduce time spent on pest control and, contrary to the tactical choices initially chosen, they applied natural plant defense stimulators and released natural enemies as preventive actions to limit the potential yield losses.

4.3. A coexistence of open-ended and deterministic perspectives on both the short-term and the long-term time scales

Farmers' experiments in Drôme Departement were described along two temporal elements: annual experiments (on a given plot and in a given year) and long-term experimental sequence (Catalogna, 2018; Catalogna et al., 2018). Both combine deterministic and open-ended perspectives.

As regards annual experiments, some can be related to a deterministic perspective. For example, "comparison" experiments consist in identifying which is the best cropping strategy among several. Several cropping strategies are implemented side by side on the same plot to compare them with one another or with the farmer's current practice. Here, farmers had specified goals to reach, they planned in advance the experimental layout to allow for comparisons and the mandatory information to gather. By contrast, other types of experiments adopt an open-ended way of experimentation, where only part of the experimental layout is decided before the onset. This is the case in "breakthrough" experiments, when the farmers broaden their inquiry on an ecological process and try it out to evaluate possible consequences on their system. For example, one farmer experimented for the first time with leaving chards to flower to breed ladybugs, wondering if they could then move into the next tomato crop and act as natural enemies of tomato pests. To check the proof of concept of this practice, he had no preconceived ideas but was open to any new information or observation that could occur during the experiment. This process rather refers to Lyon's (1996) definition of "learning during action" and Millar's (1994) definition of "adaptive experiments," while "comparison" experiments refer to more formal experiments realized by researchers to test a hypothesis (Maat and Glover, 2007; Leitgeb et al., 2014).

In the long run, most farmers' experimental sequences evolved over time from an open-ended to a more deterministic perspective (Fig. 1c, Farmer 1). They often started with an open-ended "breakthrough" experiment, to discover whether an ecological process could be activated on the farm. When promising, they would run several "improvement" experiments with little change at a time to optimize the desired ecological processes in an iterative learning loop, a process sometimes called "trial-and-error" (Lyon, 1996). Finally, they would implement a formal "comparison" experiment in which the performance of the new cropping strategies was more precisely quantified and compared to current practices, which enabled them to gain more confidence or convince peers of the interest of the practice (Lyon, 1996). But some farmers had an experimental sequence that was quite the opposite: they started in a very deterministic and reductionist approach by framing a cropping system to reach, then they split the technical problems into several singular questions. For example, a farmer experimenting conservation agriculture assumed he had to simultaneously explore three topics: direct seeding, intercropping and introduction of cover crops. Each topic was experimented on a separate field with an open-ended perspective to learn by doing (Fig. 1c, Farmer 2).

In this case study, the focus was on the individual experimental and learning processes, although farmers often exchange knowledge with peers and other stakeholders. That collective process was particularly observed in the following case study.

4.4. A combination of deterministic and open-ended perspectives relating to a separation of roles between farmers and researchers

The SOIL network, which involved both researchers and farmers, aimed to co-design and experiment with on-farm cropping systems to promote soil life. Four groups of farmers were created to address the issue in different conditions (farms devoted to grain production, to seed production, combining crops and animal husbandry, or with a specific light soil). In each group, researchers in agronomy and soil science facilitated the process to collectively design a new cropping system capable of reaching a number of prioritized objectives. For example, the "grain" group set 4 objectives for the prototype to test: direct seeding for crops and intercrops, improved soil fertility, pesticide reduction, and yield improvements. Key performance indicators associated with a satisfactory threshold and decision rules for the technical management were collectively defined. The cropping system built by the group was tested by a pilot farmer and compared to a control treatment (i.e. his current practices). This phase led to set a framework (Fig. 1d) in a deterministic way to ensure that the experiments would address a key question for each pilot farmer, and that researchers would obtain data pertinent to their research questions. Once the main decisions were framed, the day-to-day crop management was fully delegated to pilot farmers: The researchers sought to test their indicators in real cropping conditions, and held the conviction that the farmers were experts capable of managing crops in the most appropriate way from their farming context. They agreed that cropping decisions could be adapted from the decision rules initially set, as long as the deviation between what was collectively planned and what was realized by the pilot farmer could be documented. The research team took charge of the monitoring of the experiment as planned initially, not only with performance indicators co-defined with farmers but also indicators to advance scientific understanding on soil biological quality.

Deterministic vs open-ended perspectives in experimentation

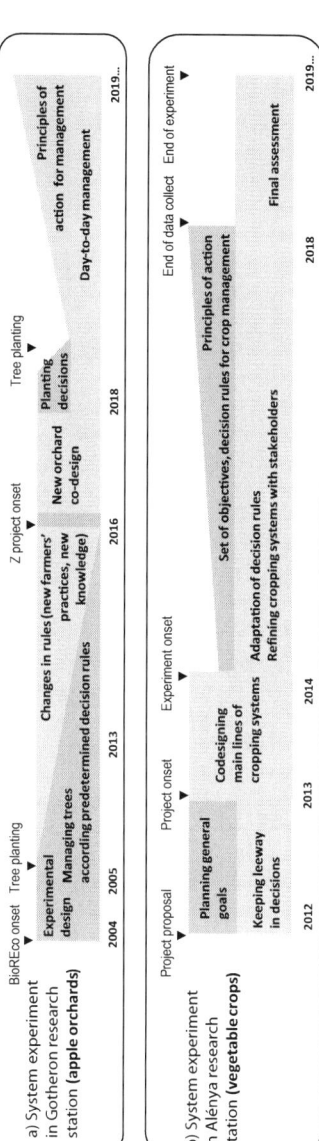

Fig. 1: Coexistence of deterministic (in blue) and open-ended (in yellow) perspectives along the experiments in the four case studies.

Such joint experiments can be related to those described by Lyon (1996) where agronomists follow up farmers' experimentation with their own formal protocols, to produce more generic knowledge. Experiments in this CS, as the three others, associated deterministic and open-ended perspectives, but here with a clear separation between researchers' and farmers' scope of decisions, and a coordination between them during collective workshops.

5. Discussion

The main finding from the cross-case analysis is that the evolving coexistence between open-ended and deterministic perspectives enables managing and learning in agroecological experimentation. After considering the different forms of such coexistence and the reasons for them (5.1), we describe some of their limits (5.2), and argue for working towards a more effective intertwining to support a large-scale agroecological transition (5.3).

5.1. Various forms of coexistence according the types of decision and over time

The challenge is to make two types of decisions interact in agroecological experiments: (i) goals to reach, means and assessment methods; and (ii) day-to-day crop management and monitoring.

With respect to goals and means for the experiments, in some CS, current available knowledge was used to draw hypotheses prior to the implementation of the experiment (as in factorial experiments) and to determine theoretical cropping systems that would be most likely to reach the desired goals (e.g., in BioREco). But in most CS that we evaluated, the experiments were not strictly framed from the beginning (4SYSLEG project, Z project), and were progressively refined during the project. The reasons were that, for a very disruptive innovation, the targets to reach could not be planned entirely; this was a major reason for allowing the open-ended and deterministic perspectives to coexist.

Three patterns of coexistence were identified in the long-term dynamics: a strict succession of the two perspectives, a progressive replacement of one perspective by the other, or the development over time of the two perspectives in parallel. The three patterns could even be present in a same project. The Gotheron CS offers an interesting example of the succession pattern, in which the same researchers adopted two opposite perspectives over 15 years: a closely framed experiment (BioREco) was later reopened to keep up with changes in commercial orchards. It was followed by the Z project, where they adopted a very open approach until trees were planted, which restrained leeway on the following crop management decisions. For some farmers in the Drôme CS, the two perspectives coexisted at the same time, with one increasing while the other one decreased (Farmer 1). And in SOIL CS, both perspectives were maintained in parallel all along the project, because each was embodied by one specific actor: researchers sought to adopt their logic of a "farm laboratory" and set minimum requirements for the experiments. But they allowed the farmers to decide on the most realistic crop management possible in an open perspective. Nevertheless, it would be a stereotype to consider that farmers would systematically experiment in a more open-ended way than researchers: Farmer 2's deterministic experiments and the scientific open-ended Z project combat this very notion.

As regards crop management decisions, for all CS, all management decisions that can reasonably be determined with a degree of confidence are implemented, but flexibility is maintained for other decisions, to adapt to climate, soil and plant conditions and to farmers' preferences in some cases. Our results highlight how experimenters embrace the issue of uncertainty on agroecological systems, either it comes from current gaps in the state of the art on agroecological regulations, from dynamics of living organisms or from farm constraints. Experimenters at research stations framed the decision process with decision rules or practical guidelines (4SYSLEG, BioREco) or monitored why changes occurred (SOIL project). Such a combination of a predetermined framework alongside flexibility can be related to adaptive management (Foxon et al., 2009). In this paradigm, "managers acknowledge the limits to predictability [...] and recognize that knowledge about social and ecological systems is both uncertain and pluralistic."

5.2. Limits to coexistence

We now discuss whether the coexistence of deterministic and open-ended perspectives could impede experimentation projects, interactions with stakeholders, and processes of knowledge building and dissemination.

In the two CS on research stations, complying with funders specifications partly forced the experimenters into a deterministic position. The public funders wanted to know, before awarding funding, the exact level of pesticide reduction and the technical means that would be engaged to achieve that goal. It was necessary to convince them of the interest of the open-ended perspective for a part of the experimental decisions.

The combination of open-ended and deterministic perspectives seemed to facilitate interactions with non-professional experimenters. At the beginning of the Z project, the design of the experimental cropping systems with agricultural stakeholders enabled brainstorming on a wide range of new ideas to build an orchard based on natural regulations and biodiversity. This participatory process within a research station was called "semi-confined experimentation" by Cardona et al. (2018), to highlight the idea that such open processes permit and value the contribution of agricultural stakeholders in the design of agroecological systems while maintaining a scientific basis of experimentation. The experiments in the SOIL project could also be related to semi-confined experimentation, with the particularity that there was here a clear separation of roles and responsibilities between researchers and farmers in the process.

Professional experimenters in 4SYSLEG and BioREco/Z projects considered it would be impossible (or at least very difficult) to learn from a constantly evolving situation. They needed a deterministic phase to frame the system depending on scientific hypotheses. Later, when some decisions were re-explored, it was critical for them to be completely aware of what phase they were in, and of the fixed specifications with which they had comply and those that could be changed. Besides, Catalogna (2018) observed situations where farmers failed to establish adequate stability over time in experimentation, and thus were consequently lost in their experimentation process and unable to learn from it.

Another difficulty relates to the dissemination of results due to open-ended processes. Researchers experimenting step-by-step designed systems had difficulties in relating the experimental protocol and the outcomes to an external audience, in both the scientific and the agricultural spheres, because of the open-ended phases. Stakeholders visiting experimental stations expected so-called "hard science based on rigorous protocols" and were surprised by experiments on evolving systems. One reason is that they are often unaware of the fact that such experiments rely on rigorous decision processes as well. This difficulty led researchers to explore in greater depth the question of decision traceability: Which experimenting decisions were taken initially? Which ones had to be changed to progressively improve the system and why? Some tools were designed to trace such decisions: a computerized table sheet in the 4SYS-LEG project, detailed reports and decision fishbones in the Z project (Penvern *et al.*, 2018). On-farm, the analytical framework proposed by Catalogna (2018) could help farmers to trace *ex post* the successive experiments on a timeline and the reasons for their sequencing. But extensive research is still required to build methods to fully synthesize decisions on open-ended experiments.

5.3. From a dual vision towards intertwining various experiments embedding open-ended and deterministic perspectives

We consider that the position of a particular experiment according to deterministic and open-ended perspectives results not only from diversity in the visions of transition to agroecology, but also from the type of question under study (e.g., more or less systemic) and the position according to knowledge (e.g., stabilizing knowledge on a specific question or exploring new ones). A major outcome of our study is a deeper characterization of each experimenting phase. An open-ended vision of agroecology supports the idea that uncertainty and unpredictability are inherent to agroecological systems; it emphasizes improvement loops, multi-actor exchanges, learning from ecological processes and stakeholders' points of view (Altieri, 2002; Francis, 2003; Cristofari et al., 2018). For experimentation, it corresponds to explorative periods where new learning occurs in relation to real-time ecosystem reactions and stakeholder exchanges. Nevertheless, an open-ended vision of the agroecological transition does not completely discredit deterministic experimentation. Experimenting in a deterministic way matches a vision of agroecology where the system to manage is considered as complex but predictable, and where the available knowledge is sufficient to predict the probable effects in advance, therefore where the biophysical system can be steered by humans. For experimentation, a deterministic phase corresponds to a willingness to control a certain number of parameters in order to make a proof, settle controversies and hence stabilize certain knowledge. This was particularly visible when experimentation was steered by researchers, who regularly closed some avenues of exploration and deepened others.

Far from viewing the different ways of experimenting in opposition to one another, we argue for enabling a tight and explicit intertwining between them. In the previous case studies, the combination of both perspectives rather came along the way – by default – as problems emerged or socio-technical context evolved. It is now necessary to conceptualize the intertwining between both, based on agronomy and agroecology theories and convince other scientific or non-scientific actors of its interest. The challenge is to make more explicit the scientific reasoning of what sort of experiments to implement depending whether the aim is to stabilize existing knowledge or explore new avenues, to learn new knowledge, disseminate the acquired one, or confront different knowledge. The challenge for agronomy researchers is also operational, to help all experimenters, whoever they are, to better clarify why, how and when they combine the two perspectives, and to invent tools to favor such exploration, in particular to trace decision making, to collect relevant data and to assess the performance of the systems. A significant issue for the future of agroecological experimentation consists in our capacity to gather and analyze multi-local and multi-actor data to support a large-scale agroecological transition. It is mandatory to build tools with advisory services to support knowledge exchanges among groups of farmers and favor learning. It is also mandatory to build alternative statistical methods, capable to take advantage of the rich but heterogeneous qualitative assessment from stakeholders.

6. Conclusion

This chapter proposed a cross-case analysis of four experimentation projects differing in several respects: the experimental sites (on farm or research station); the actors and their degree of involvement; the objects under study; and the intensity with which ecological processes were mobilized. We demonstrated that the combinations between deterministic and open-ended experimentations varied, between the actors involved, the periods of the projects, and the decisions made. The coexistence of the two perspectives appeared as an operational way to address the issue of uncertainty on the ecological process and could renew experimental methods. Despite prominent technical aspects in agronomy, the present analysis also highlights social change involved in the agroecological transition where knowledge and experiences are shared between scientific and non-scientific experts. Agronomy could be enriched by acknowledging a range of intertwining experimental approaches, both from a scientific perspective and from an action perspective, to accompany farmers in the agroecological transition.

Acknowledgements

The authors thank the following funders: PSDR4 Rhône-Alpes (Pour et Sur le Développement Régional) for funding the COTRAE project who benefited from fundings from INRAE, Auvergne-Rhône-Alpes Region, and European Union (FEADER, European Innovation Partnership EIP-AGRI); Ecophyto (4SYSLEG and BIORECO/Z projects), and Drôme Departement (Catalogna's PhD Thesis). They also thank Liz Libbrecht for reviewing the English language.

References

Alaphilippe, A., Simon, S., Brun, L., Hayer, F., Gaillard, G., 2013. Life Cycle Analysis Reveals Higher Agroecological Benefits of Organic and Low-Input Apple Production, *Agronomy for Sustainable Development*, 33, 3, 581–592.

Altieri, M. A., 2002. Agroecology: The Science of Natural Resource Management for Poor Farmers in Marginal Environments, *Agriculture, Ecosystems and Environment*, 1971, 1–24.

Berthet, E. T. A., Barnaud, C., Girard, N., Labatut, J., Martin, G., 2016. How to Foster Agroecological Innovations? A Comparison of Participatory Design Methods, *Journal of Environmental Planning and Management*, 59, 2, 280–301.

Boidron, A., 2018. *Rôle de la recherche participative dans l'accompagnement des agriculteurs vers l'agroécologie. Étude de cas: le Collectif Sol*. Mémoire fin d'étude, ISA, Lille.

Cardona, A., Lefèvre, A., Simon, S., 2018. Les stations expérimentales comme lieux de production des savoirs agronomiques semi-confinés. Enquête dans deux stations INRAE engagées dans l'agro-écologie, *Revue d'anthropologie des connaissances*, 12, 2, 139–170.

Catalogna, M., 2018. *Expérimentations de pratiques agroécologiques réalisées par les agriculteurs. Proposition d'un cadre d'analyse à partir du cas des grandes cultures et du maraîchage diversifié dans le département de la Drôme*, PhD thesis, Avignon.

Catalogna, M., Dubois, M., Navarrete, M., 2018. Diversity of Experimentation by Farmers Engaged in Agroecology, *Agronomy for Sustainable Development*, 38, 5, 50.

Coquil, X., Fiorelli, J.-L., Blouet, A., Mignolet, C., 2014. Experiencing Organic Mixed Crop Dairy Systems: A Step-by-Step Design Centred on a Long-Term Experiment, in Bellon, S., Penvern, S. (Eds.), *Organic Farming, Prototype for Sustainable Agricultures*, Dordrecht, Springer Netherlands, 201–217.

Cristofari, H., Girard, N., Magda, D., 2018. How Agroecological Farmers Develop Their Own Practices: A Framework to Describe Their Learning Processes, *Agroecology and Sustainable Food Systems*, 42, 7, 777–795.

Debaeke, P., Munier-Jolain, N., Bertrand, M., Guichard, L., Nolot, J. M., Faloya, V., Saulas, P., 2009. Iterative Design and Evaluation of Rule Based Cropping Systems: Methodology and Case Studies. A Review, *Agronomy for Sustainable Development*, 29, 1, 73–86.

Doré, T., Makowski, D., Malézieux, E., Munier-Jolain, N., Tchamitchian, M., Tittonell, P., 2011. Facing up to the Paradigm of Ecological Intensification in Agronomy: Revisiting Methods, Concepts and Knowledge, *European Journal of Agronomy*, 34, 4, 197–210.

Duru, M., Therond, O., Fares, M., 2015. Designing Agroecological Transitions; A Review, *Agronomy for Sustainable Development*, 35, 4, 1237–1257.

Foxon, T. J., Reed, M. S., Stringer, L. C., 2009. Governing Long-Term Social–Ecological Change: What Can the Adaptive Management and Transition Management Approaches Learn from Each Other? *Environmental Policy and Governance*, 19, 3–20.

Francis, C., Lieblein, G., Gliessman, S., Breland, T. A., Creamer, N., Harwood, R., Salomonsson, L., Helenius, J., Rickerl, D., Salvador, R., Wiedenhoeft, M., Simmons, S., Allen, P., Altieri, M., Flora, C., Poincelot, R.,

2003. Agroecology: The Ecology of Food Systems, *Journal of Sustainable Agriculture*, 22, 3, 99–118.

Johnson, A. W., 1972. Individuality and Experimentation in Traditional Agriculture, *Human Ecology*, 1, 2, 149–159.

Klerkx, L., Leeuwis, C., 2008. Balancing Multiple Interests: Embedding Innovation Intermediation in the Agricultural Knowledge Infrastructure, *Technovation*, 28, 6, 364–378.

Lechenet, M., Deytieux, V., Antichi, D., Aubertot, J. N., Bàrberi, P., Bertrand, B., Cellier, V., Charles, R., Colnenne-David, C., Dachbrodt-Saaydeh, S., Debaeke, Ph., Doré, T., Farcy, P., Fernandez-Quintanilla, C., Grandeau, G., Hawes, C., Jouy, L., Justes, E., Kierzek, R., Kudsk, P., Ram Lamichhane, J., Lescourret, F., Mazzoncini, M., Melander, B., Messéan, A., Moonen, A. C., Newton, A. C., Nolot, J. M., Panozzo, S., Retaureau, P., Sattin, M., Schwarz, J., Toqué, C., Vasileiadis, V. P., Munier-Jolain, N., 2017. Diversity of Methodologies to Experiment Integrated Pest Management in Arable Cropping Systems: Analysis and Reflections Based on a European Network, *European Journal of Agronomy*, 83, 86–99.

Lefèvre, A., Salembier, C., Perrin, B., Lesur-Dumoulin, C., Meynard, J. M., 2015. Design, Experimentation and Assessment of Four Protected Vegetable Cropping Systems Adapted to Different Food Systems, in *Proceedings 5th International Symposium for Farming Systems Design*, Montpellier, France, September 7–10.

Leitgeb, F., Kummer, S., Funes-Monzote, F. R., Vogl, C. R., 2014. Farmers' Experiments in Cuba, *Renewable Agriculture and Food Systems*, 29, 1, 48–64.

Lyon, F., 1996. How Farmers Research and Learn: The Case of Arable Farmers of East Anglia, UK, *Agriculture and Human Values*, 13, 4, 39–47.

Maat, H., Glover, D., 2007. Alternative Configurations of Agronomic Experimentation, in Sumberg, J., Thompson, J. (Eds.), *Contested Agronomy: Agricultural Research in a Changing World*, New-York, Routledge, 131–145.

Meynard, J. M., Dedieu, B., Bos, A. P., 2012. Re-Design and Co-Design of Farming Systems. An Overview of Methods and Practices, in Darnhofer, I., Gibbon, D., Dedieu, B. (Eds.), *Farming Systems Research into the 21st Century: The New Dynamic*, Netherlands, Springer, 407–432.

Millar, D., 1994. Experimenting Farmers in Northern Ghana, in Scoones, I., Thompson, J. (Eds.), *Beyond Farmer First. Rural Peoples' Knowledge,*

Agricultural Research and Extension Practice, London, ITDG Publishing, 160–165.

Navarrete, M., Lefèvre, A., Dufils, A., Parès, L., Perrin, B., 2017. Concevoir et évaluer avec les acteurs des systèmes de culture adaptés à leurs cadres de contraintes et d'objectifs en production maraichère sous abri. Mise en pratique et enseignements dans les projets GeDuNem et 4SYSLEG, *Innovations Agronomiques*, 61, 33–49.

Navarrete, M., Brives, H., Catalogna, M., Gouttenoire, L., Heinisch, C., Lamine, C., Ollion, E., Simon, S., 2018. Farmers' Involvement in Collective Experimental Designs in a French Region, Rhône-Alpes. How Do They Contribute to Farmers' Learning and Facilitate the Agroecological Transition? *in Proceedings 13th International Symposium for Farming Systems Association*, Chania, Greece, July 1–5.

Penvern, S., Chieze, B., Simon, S., 2018. Trade-Offs Between Dreams and Reality: Agroecological Orchard Co-Design, in *Proceedings 13th International Symposium for Farming Systems Association*, Chania, Greece, July 1–5.

Perrin, B., Lefèvre, A., 2019. L'association culturale, un levier pour améliorer santé des plantes, fertilité du sol et production des systèmes de culture maraichers diversifiés? *Innovations Agronomiques*, 76, 51–70.

Prost, L., Berthet, E. T. A., Cerf, M., Jeuffroy, M. H., Labatut, J., Meynard, J. M., 2016. Innovative Design for Agriculture in the Move towards Sustainability: Scientific Challenges. *Research in Engineering Design*, 28, 1, 119–129.

Saad, N., 2002. *Farmer Processes of Experimentation and Innovation: A Review of the Literature*, Report CGIAR systemwide program on participatory research and gender analysis, Cali.

Salembier, C., Segrestin, B., Berthet, E., Weil, B., Meynard, J. M., 2018. Genealogy of Design Reasoning in Agronomy: Lessons for Supporting the Design of Agricultural Systems, *Agricultural Systems*, 164, 277–290.

Simon, S., Brun, L., Guinaudeau, J., Sauphanor, B., 2011. Pesticide Use in Current and Innovative Apple Orchard Systems, *Agronomy for Sustainable Development*, 31, 3, 541–555.

Plant breeding for agroecology: A sociological analysis of the co-creation of varieties and the collectives involved

SOPHIE TABOURET, CLAIRE LAMINE,
FRANÇOIS HOCHEREAU

1. Introduction

Agroecology is a concept that progressively entered French agricultural research, networks and policies in the last decade, generating a diversity of definitions and controversies (Bellon and Ollivier, 2018; Compagnone et al., 2018). In the early 2010s, INRA[1]'s Genetics Department started to work on its application to plant breeding, based on various techniques and on new ways of working with agricultural actors in order to reduce the negative impacts of agriculture on its environment (Litrico et al., 2014). Scientists in biology and plant breeding have produced different concepts to define the variability of living organisms and to qualify varieties (of cultivated plants)[2]. It includes the genotype (a part of an individual's genetic information), the phenotype (the set of observable traits, whether they are linked or not with genes), a concept introduced by Bateson at the beginning of the 20th century, and the ideotype, a concept proposed by Donald (Donald, 1968) to consider the variety based not only on isolated characteristics such as defect or yield but also on the agronomic conditions of growing. This last definition of ideotype

---

[1] INRA is the National Institute for Agronomic Research in France, and also the main breeder in vines and apricot trees in France. It became INRAE in 2020 through the merger of INRA and IRSTEA.

[2] The term "cultivated plant varieties" must be understood here in its administrative acceptance. The varieties are registered in the official catalogue to be grown in France (https://www.geves.fr/catalogue/ accessed on the 31st of October 2019).

has then been extended in order to refer to the expected characteristics in relation to specific environments and production methods (Debaeke et al., 2014). The ideotype is a concept brought by natural sciences, and as we will see, it is an interesting object for social sciences.

In the social sciences, the literature about plant breeding is still sparse. Most recent works have studied alternative farmers movements which work mainly on population seeds for peasant agriculture (Demeulenaere, 2014; Coolsaet, 2016; Derbez, 2018) with the exception of a few studies on more conventional commodity chains such as the fruit sector (Vanloqueren and Baret, 2008; Lamine, 2017). When the concept of "ideotype" is mentioned, it is quoted as an emic and technical term (Lamine, 2014; Belmin et al., 2018). Crossing genetics and sociological perspectives, some authors have discussed this concept of ideotype for the case of annual plants in organic production (Desclaux et al., 2013). They have shown that it is not relevant for annual plant breeding in organic agriculture because it is mainly focused on the question of yield and anchored in a productivist paradigm, in which the use of inputs (fertilizers and plant protection products) allows a maximization of harvests (Lamine et al., 2011). To support input reduction, varietal innovation has to consider the diversity of production contexts (pedoclimatic and human environment) in a dynamic way. These authors thus proposed the concept of realtype to adapt the ideotype to field reality. Their main contribution is to highlight the importance, in the success of the variety, of relevant collectives that can take over the problems appearing with the changes in the context (Desclaux et al., 2013). This notion of relevant collectives points out the need to involve the various expertizes able to incorporate the diversity and the evolution of the contexts of use of a future variety. In organic agriculture (Desclaux et al., study is on organic agriculture but we think it can be useful for agroecology as well), the diversification of environments requires that genetic innovation be both dynamic (able to evolve over time) and participatory (involving geneticists *and* stakeholders anchored in these environments) (Sylvander et al., 2006). In the view of these authors, this also allows to take into account the increasing uncertainties faced by the agricultural world. The article by Desclaux and al. has often been cited but little discussed and we believe it to be important to articulate the two notions of "relevant collective" and "uncertainty" which will be key to our analysis.

We will discuss two case studies related to varietal creation in grapevine, and one case study related to varietal creation in apricot. All these cases involve perennial plants (Vanloqueren and Baret, 2004; Lamine et al., 2015). This context slightly modifies the notions of relevant collective and uncertainty in relation to the way they were approached in relation to annual plants by Desclaux et al. Beyond the lengthy time taken to create a variety (about twenty years), there is also the fact that perennial plants cannot be uprooted and replaced every year to adapt to market conditions or to overcome health problems. Working over the long time period of perennial plant breeding means working in a changing world where the variety created will be part of a very different sociotechnical context from the one in which it was conceived. Uncertainty regarding the ability to create a variety adapted to the problems of the actors is thus much stronger. The cases that we are observing shake up the acceptance of what should be a relevant collective, jointly associating farmers, geneticists and even other actors. All of the issues addressed by the various stakeholders: taste, resistance to disease, market, adaptation to terroirs etc. are all elements that involve a high degree of uncertainty. With regard to taste, some authors even go as far as to show that the product tasted and the taster are shaped, or even transformed, by each other (Hennion, 2005; Méadel and Rabeharisoa, 2013). This chapter aims to observe how relevant collectives emerge in this case of perennial plants, how they are composed, how they deal with these uncertainties. We will show that along the process of creation of appropriate varieties for agroecology, these collectives adopt or articulate more or less openended or deterministic perspectives over time.

Indeed, the agroecological transition in varietal innovation implies a transition from a "fixist paradigm" in which "appropriate varieties" with predictable and stable behaviour and high potential yields can and should be created (Bonneuil and Hochereau, 2008; Bonneuil and Thomas, 2009) – which relates to the notion of ideotype – to an "open ended" perspective embodied by a more participative approach that integrates the uncertainties. This latter perspective is based on the principle of increasing the capacity of anticipation of actors involved in plant breeding and allows the integration of the variety "in the making" into a socio-technical network that evolves in a flexible way in response to the problems and new uncertainties encountered. The definition of an "appropriate variety" according to standardized criteria, regardless of the contexts of use, no longer holds in the face of the variability of environments, the diversity of practices in localized agricultural systems and the growing demands of society for an ecological agriculture.

In this chapter, the goal is to overcome the duality between, on one hand, a positivist vision of things and processes whose characteristics we must reveal and, on the other hand, a vision of things that are pure social constructions (Callon, 1986; Bessy and Chateauraynaud, 1993). On the contrary, we think that varieties and the collectives that shape them are produced together and by each other. Varieties emerge not only from spaces where cross breeding are realized, but also through experimentation, relationships with diseases, markets, terroirs and so on. As we will see in our cases studies, collectives emerge (and transform) around these varieties in the making. Our approach will describe the emergence and promises of new collectives – some of which we have initiated and tested – that involve a diversity of stakeholders representing the different components of the socio-technical system. We posit that a conceptualization of their co-presence is a way to tackle the interdependencies that link them and block agroecological transitions in genetic innovation (Lamine, 2017) as well as a way to collectively face uncertainties.

2. Three case studies in perennial plant breeding

Based on three case studies, we study the socio-technical networks in which a consensus on the appropriate ideotype emerges. Plant breeding is an axis of innovation that has already proved its importance during sanitary crises in wine (many hybridisations were conducted after the late 19th century Phylloxera crisis) and fruit production (*sharka* for the apricot tree). Disease-resistant varieties then proved to be effective solutions. They also make it possible to reduce or suppress pesticide use (Hochereau et al., 2015), and thus meet societal expectations of an ecological agriculture, be it associated to organic agriculture, to agroecology or to other models. This has led geneticists to work more closely with professionals in order to define ideotypes adapted to different contexts (Litrico et al., 2014). Two main elements allow us to distinguish these three cases studies from each other (Tab. 1). The first one is the step of the creation process when the project takes place. in the Languedoc wine case, varieties are already evaluated while hybridisations[3] are in progress in the Rosé de Provence's case and not yet started in the Prunus one. The second difference deals with the actors involved in the process. In

---

[3] Hybridization is a sexual reproduction between two different but closed species.

Tab. 1: Characterization of the three case studies.

	Languedoc Wine case	Rosé de Provence	Prunus
State of the creation process	Hybridizations finished, some genotypes are evaluated.	Hybridizations in progress	Before hybridizations
Actors involved	– scientists (wine making process, phytopathologists) – agricultural technicians – interbranch organization	– oenologists, breeders – PDO management bodies – winemakers – producers' union - marketers	– researchers – experimenters – producers – advisors – breeders – nurseries
Focus of the analysis	Tracing a controversy about resistance's sustainability.	Identifying what links varietal innovation to local conditions of uses and terroirs.	Co-building a systemic way to manage the agroecological design of an ideotype.

all three cases, researchers are involved, whether they are oenologists, breeders or economists and so on. Producers are present or represented by their unions and technicians and different actors of the supply chain are involved in each case.

Participant observation and interviews with concerned actors are our main sources of data. Our foci of analysis are quite different, in link with the different role of the sociologist in each case study, but all allowing us to tackle the co-creation of varieties and their relevant collectives.

2.1. Languedoc Wine: New actors that change the definition of a "sustainable" resistance[4]

In the late 1990s and early 2000s, INRA – which is also the leading breeder of grape varieties in France – took a new orientation in varietal

---

[4] Between September 2015 and August 2018, Sophie Tabouret conducted sixteen interviews with INRA researchers and actors from the Languedoc region mobilized to experiment with Bouquet varieties, took part in twelve meetings about their experimentation and one tasting for Languedoc wine growers. She also reviewed the scientific literature related to the durability of resistance to mildew and powdery mildew of vine varieties and thirteen press articles on the Bouquet varieties published in *Vitisphère*.

selection in vines by working on sustainable resistant varieties. The ResDur program[5] is based on preliminary work conducted since the 1970s by Alain Bouquet, a geneticist who selected many varieties with one resistant gene against downy mildew and another one against powdery mildew. ResDur varieties have inherited the resistance genes already present in Bouquet varieties, and other resistance genes have been added. The idea is that a pathogen will have even more difficulty bypassing the resistance of a variety if there are several resistance genes, rather than only one[6]. This led INRA to disregard the previously created Bouquet varieties as presenting unsustainable resistance, and to refuse to consider their inclusion in the official catalogue.

However, Bouquet's varieties have been successfully tested for several years as part of a local research program conducted by the Languedoc wine industry. Since the beginning of 2010, many Languedoc actors (oenologists, winegrowers or technicians) have been gathering around Bouquet varieties to make them available for production. The quality of the wines is also promising but INRA's research direction refused to experiment these varieties for a commercial use. INRA's decision to stop experimenting with Bouquet's varieties, in the name of a precautionary principle to protect resistance genes, was difficult to accept by local professionals. They highlighted the good adaptation of these grape varieties to the Languedoc pedoclimatic context, but also a good correspondence between the wines produced with these varieties and the typicity of Languedoc wines (fruity). Finally, they insisted on the very high varietal resistance of these varieties to the pressure of local pathogens, which would allow them to respond quickly to the societal injunction to reduce pesticides without reducing the quality of their wines. According to them, the hybridisations carried out for the ResDur program did not allow the same level of quality to be achieved and their adaptation to the Languedoc context would still require many years. The institute finally reversed its decision concerning Bouquet varieties and accepted an experimentation at the beginning of 2017. With this case study, we observe a shift from a technicist definition of resistant varieties

---

[5] ResDur for « Résistance Durable » (Sustainable Resistance in French).
[6] Delmas, C. E. L., F. Fabre, J. Jolivet, I. Mazet, S. Richart Cervera, L. Delière, et F. Delmotte. 2016. « Adaptation of a plant pathogen to partial host resistance: selection for greater aggressiveness in grapevine downy mildew. » *Evolutionary Applications* 9: pp. 709–725.

to a definition that appears more open to the experience of regional winegrowers whose practices change the definition of resistant varieties.

This controversy crystallises both around the uncertainties related to the (un)sustainability of resistance, and the willingness of stakeholders to provide new answers and definitions of "sustainable resistance." It also relates to several fundamental tensions on what makes a sustainable resistance, between geneticists and oenologists but also between geneticists' experiments and field experience of wine professionals. Tracing the controversy has allowed us to see the emergence of actors, whether they are scientists, professionals, as well as genes, varieties, pathogen's populations and their networks.

At the beginning of the controversy (in the early 2010s), there was no doubt for the direction of INRA that the Bouquet varieties were condemned because of their monogenic resistance. ResDur varieties had been created to overcome this weakness. However, scientific studies on resistance breakdowns in phytopathology and genetics showed growing uncertainties about the fragility of Bouquet varieties' resistance. The initially structuring opposition between monogenic resistance varieties (such as supposedly Bouquet varieties) and polygenic ones (such as ResDur varieties) gradually broke down. The results of a study published in 2016 showed that mildew harvested on resistant varieties was more aggressive than that harvested on sensitive varieties. This showed an adaptation of the pathogen, which creates an erosion of partial resistance[7]. This work suggested the need to consider the deployment of resistant varieties combined with agronomic surveillance of resistance through biological control or fungicide. The identification of a possible downy mildew able to break down the genetic resistance, led to propose an experimental protocol to follow resistant varieties and their pathogens. The winegrower thus becomes an ally to protect the resistance of the vine but also to better adapt varieties to different contexts of use.

A change in posture, which is reflected even within INRA's management, is thus taking place[8]. Monitoring pathogens, thinking in terms of their characteristics and means of action, leads researchers to consider

---

[7] See note 6.
[8] In 2020, 4 INRAE-Resdur varieties obtained a "definitive registration" and can be grown whitout any restriction whereas 11 INRAE-Bouquet ones obtained a "temporary registration" and can be grown under an experimental protocol (http://observatoire-cepages-resistants.fr/documentation/reglementation/ accessed on the 12nd of November 2020).

other modes of defence than the accumulation of genetic barriers. They suggest reinforcing the varietal resistance with the use of fungicides at key periods in order to support the sustainability of the resistance. This case study suggests that the potential of a plant to resist an aggressor becomes a property of the variety anchored in a socio-technical environment. It includes pathogen populations, genes and fungicides, etc., and is no longer a property only defined by the number of resistance genes.

This case study sheds light on a movement that starts with a resistant varietal ideotype. The way to obtain varieties with a sustainable resistance to downy mildew and powdery mildew is at first deterministic: only varieties based on pyramiding of several powdery and downy mildew resistant genes produce a sustainable resistance to breeders from INRA. The controversy questions this vision. The variety is part of a socio-technical network that requires us to revise this deterministic vision of the scope of agroecological transition and include new actors (winegrowers, oenologists, a more virulent pathogen, pathologists, etc.) into this socio-technical network.

2.2. Rosé de Provence: Taking into consideration the practices of concerned actors

The Centre du Rosé (CdR)[9] – an experimental centre created in 1999 and mainly financed by the interbranch organization (Conseil Interprofessionnel des Vins de Provence – CIVP) – is involved in a downy mildew and powdery mildew resistant breeding program for Rosé[10] wines from the Provence and Mediterranean region. A specialization in rosé is the main characteristic of the Provence wine production area (42% of the rosé's national production). The CdR carries out both winemaking and oenological research. Its first researches focused on winemaking techniques, which gave Rosé wines from Provence their particular colour. Today, Rosé de Provence can be described in this way:

---

[9] Between April 2016 and April 2018, Sophie Tabouret conducted seven interviews with CdR employees, associated structures and geneticists working on plant breeding for Rosé wines, took part in four tastings at the CdR, a presentation by the CdR oenologist at a wine fair. She also mobilized grey literature combining the analysis of the websites of the organisms involved in Rosés de Provence's marketing and articles published by the CdR.

[10] This capital letter is used on all documents produced by CdR. We decide to use it with the capital letter R as well.

"*Very pale Rosé wines, very light colour and therefore aromatic, very aromatic and this aroma is on the fruity side. There are some floral touches that can arise.*"[11]

Two major sensory characteristics emerge from this description: pale colour and fruity aromas. The project of varietal creation starts with the definition of ideotypes that would be adapted for Rosé wine in Provence. Breeders are arguing for a definition of ideotypes based on the oenological knowledge developed by CdR.

Even though the above description reflects the work done to improve the quality of Rosés de Provence wines, it is also problematic. The quality of these wines is now appreciated thanks to the work carried out on winemaking techniques, an overall improvement in the quality of Rosés from other French areas has also been observed and a risk of uniformity of rosé wines regardless the origin of the grapes appears. It also reinforces the vision of an industrial wine: "*we understood that if we used a lot of technical expertise in the cellar, there was the risk of suggesting that this wine was an industrial product. Mastered by cold, stainless steel and a science of engineers in white coats.*"[12] It thus becomes necessary to invest other aspects of wine production than just winemaking techniques. As in the first case study, the innovation is to propose "[resistant] grape varieties adapted to the terroirs" (Delière et al., 2017). The postulate associated with this varietal creation project is paradoxical[13]: terroir requires adapted varieties, even though the varieties are designed to behave in the same way whatever the terroir (Bonneuil and Thomas, 2009). The link between terroir and wine quality is often construed in a romantic way, targeting an element of the physical environment to explain a particularity (Gade, 2004). Through the observation of the handling of this problem by the CdR oenologists and especially of their work on the search for new descriptors, we observe an important work to rethink the technical and social problems together.

A reflexion about the tools and practices within the CdR was developed, starting from an experiment about the typicity of French rosé wines. Several wines seen as typical by the experts of diverse

---

[11] Oenologist, CdR, September 2016, Vidauban (France).

[12] Director, CdR, April 2016, Vidauban (France).

[13] For a similar reflection on potatoes and on clementines, see (Garçon, 2015; Belmin et al., 2018)

regions of production (Provence, Languedoc, Bordeaux, Val de Loire, Burgundy and Côtes du Rhône) were tasted by the CdR professional jury. The results of this survey, presented during the first tasting on the typicality of wines in 2017, showed that there was indeed a capacity of professionals to recognize Provence wines compared to other regions' wines. Rosés de Provence – chosen for their typicity – are widely considered among the most typical wines. However, an unexpected phenomenon occurred, when the jury produced an identical description of two wines from two regions (Provence and Côtes du Rhône). This result suggested a lack in the description tools and led the CdR team to consider the use of new descriptors. Speaking of the wines of La Londe[14], an oenologist described this search for descriptors:

> *"How do we describe them? We can have fun describing them like we normally do, without words or a quantifying scale, and beyond that we can also have vocabulary that allows us to let loose, to go further and say: "the La Londe are 'whittled'". "Whittled" doesn't mean anything by itself. But if we say that we find the La Londe's to be whittled, that one of us has this good idea and that therefore we agree to say "the La Londe's are whittled", well then they're whittled. But you know, we have to come up with the thing!"* [15]

This oenologist expresses her willingness to innovate by creating new descriptors for these wines: tasting to find the right words. The classical divide between realism and constructivism reaches a climax when it comes to qualifying things. Either we consider the wine as something that has intrinsic properties that the expert must reveal, or we consider that the characteristics of the objects are socially constructed (Teil, 2009). The actors show us a third way here, and work these two ideas together by associating the perception of the typicity of wine with the production of a new vocabulary. The oenologist also combines sensory analysis with a study of winemakers' practices that can have an impact on the taste of the wine. The challenge is to take seriously the sensitive experiences and the complexity of the links actors maintain with forms of expression and judgement. There is both the need to discover intrinsic characteristics that have not been discovered yet, whether they are molecules, sensations for the taster, etc., but also the will to imagine a new vocabulary that will build the recognition of these wines. Producing

---

[14] A designation of terroir of the PDO Côtes-de-Provence.
[15] Oenologist, CdR, April 2017, Vidauban (Var).

new descriptors also expresses a will to change the practice of tasting and at the same time to change the wine tasted. To name a taste, a smell, is to take a step in the relationship with the object being tasted.

In addition to defining varietal ideotypes with geneticists to promote the expected characteristics, the CdR is investing to create new links with Rosés de Provence wines through the tasting networks it has put in place. Tasting is the CdR's central working tool, and offers another way to explore and define future varieties. These concerned actors are for the first time involved in a vine breeding program and it is through their long-term work that the varietal ideotype for the creation of resistant grape varieties is defined.

2.2. *Prunus*: A multi-actor process to open and discuss the list of relevant criteria

The peach and apricot industry, like the fruit sector in general, is characterized by a high international competition and strong marketing constraints that are hardly compatible with environmental challenges and societal expectations (Lamine, 2017). For example, retailers impose diverse constraints regarding fruits' appearance and storage properties, which in turn are only achievable with input intensive agricultural practices. In the middle Rhône Valley (Drôme and Ardèche French Departments), the economic context of repeated and then structural crises since the early 1990s combined with sanitary problems[16] caused a sharp decrease in peach cultivated surface areas and production, and a partial substitution of peach by apricot in some farms. Until recently the breeding strategies have mainly focused on improving yield and fruit technological quality for packaging, storage and transport, and on creating "varietal series" in order to offer retailers continuity in each product type over a production period that has been extended from three to nearly six months for peach production. These objectives lead to the creation of cultivars and then market fruits that are easy to store and transport, but often disappointing in terms of taste and nutritional quality and poorly adapted to low-input practices and environmental criteria, especially for late harvesting cultivars (even if substantial work has been done

---

[16] Such as the development of the *sharka*, a disease due to a quarantaine virus which led to uproot some infested trees of the peach orchard (although its actual direct impact was in fact quite limited).

for decades on resistant cultivars). Producers are obliged to follow the turn-over and plant cultivars that allow them to remain on the market.

The objective of the Prunus and Ardu projects (2013–2016) was to address this key question: How can fruit production become more ecological, considering the scale of the regional socio-technical system of fruit production and the interdependencies between its actors? An interdisciplinary team (sociologists, geneticists, agronomists, economists, geographers) was established, along with a steering group of about ten people, gathering these researchers, farmers, advisers, and market intermediaries. In a first step the research team, in close interaction with this reflection group, carried out a socio-historical analysis of the evolution of the regional fruit chain (socio-technical system) based on focused surveys and documentary analysis as well as an analysis of producers' practices and strategies. This showed that the trajectory of the socio-technical system of the peach and apricot sectors was determined by a range of interconnected interdependencies involving public land planning, farming strategies, advisory system, organization of marketing, breeders' strategies etc. that generated lock-in effects. This trajectory translated in a shift from a genetic innovation model based on objectives and criteria linked to the production stage (regularity, yield) to an innovation model based on the marketing stage and the characteristics of the product (sugar content, acidity rate, appearance and firmness) and its aptitude for transporting and storing (post-harvest behaviour) (Lamine et al., 2015). The challenge of the Prunus project was then to re-open the list of objectives and criteria to take into account in breeding strategies.

In this aim, a cycle of three workshops devoted to the co-conception of future sustainable fruit cultivars was organized. For these workshops, the initial steering group was extended to a dozen more people who embodied the other components of the genetic innovation system (nursery operators, geneticists, evaluators, breeders, seed regulatory institutions). A shared problem was defined with the participants: What should be the criteria for sustainable fruit cultivars and how should the innovation system be redefined to favour them? In the first workshop, a large range of criteria was collectively established, discussed and prioritized. The discussion led to a reopening of the criteria to take into account and included issues such as adaptation to climate change, to low-input or organic agriculture, to a diversity of marketing outlets, or even work organization (the working time devoted to fruit thinning could be reduced if a criteria was added aimed at reducing the need for it). Like in the previous cases, the issue of polygenic versus monogenic resistance was discussed in regard to the main disease tackled by recent genetic research (the *sharka* disease in this case), but it is mainly the notion of "overall hardiness" (*rusticité globale*) that was put forward by the participants, as the trees and fruits are affected by other diseases. Within the group, some participants claimed a larger perspective on varieties resistance, taking into account the interactions with the environment and practices. Of course, actors' visions were not homogeneous and the aim was not to reach consensus but rather to leave the diversity of visions and possible trajectories open. For example, the criteria of the fruit's colour – recent selection processes have led to a favouring of red apricots for example – was considered "cosmetic" by some participants and "strategic" (in terms of product segmentation) for others.

In a second workshop, a prioritization of these traits was carried out according to two main contrasted scenarios for future fruit production systems (including marketing and consumption issues). Both scenarios shared a common orientation towards the reduction or absence of chemical inputs, and had been designed based on the analyses carried out in the first phase of the project. For each scenario, the criteria and traits were discussed and translated into description of ideotypes, that is, varieties adapted to the future challenges that had previously been identified, which were represented through radars. One of the key outcomes of the discussion was the shared acknowledgement of the need not only to enlarge the list of criteria but also to articulate genetic innovation with other innovations allowing for more ecological practices: "*Genetic innovation cannot solve everything, but when we start it, we need to also conceive other innovations, it has to take place in an integrated innovation system,*" as one participant put it. "*The choice of varieties questions farming systems, the links to the territory and an integrated strategy from the plant to the product,*" said another. Finally, during the last workshop, the necessary reconfigurations of the evaluation system were discussed though an incremental and a "de novo" (ground zero) conception perspective, which led to highlighting the need for a more coordinated (among actors) and more (spatially) distributed system based on local farmers' networks, in order to take into account the diversity of stakes and contexts. The outcome of this overall approach was to identify conditions for a cluster based breeding and evaluation process, targeting the most relevant traits for the future. Of course, it is a first step of a still ongoing process.

The collective and iterative approach related here illustrates the combination of (rather) deterministic and (rather) open-ended perspectives.

On the one hand, the methodological steps of the different seminars were conceived in deterministic ways, although they were slightly adjusted along the way. On the other hand, the participatory work aimed at identifying a large range of criteria and traits, and widely exploring possible scenarios, adopted an open-ended perspective. This approach required an enlargement of the socio-technical network from the geneticist themselves to a diversity of actors embodying the components of the innovation socio-technical system (nursery operators, geneticists, evaluators, breeders, seed regulatory institutions, advisors, farmers, and of course other disciplines – whereas some of them, such as social scientists, advisors and farmers are not involved in conventional breeding programs[17]), although the civil society and the consumers as such were not involved.

3. Discussion

Agroecology has a different acceptance in the three cases. In the context of apricots, the projects considered here were launched following a major sectorial crisis that articulated environmental, economic and social issues. In the wine industry, they were implemented following the rise of criticism of the massive use of pesticides in viticulture. Agroecology appears through disease resistance in the creation of new varieties in viticulture, while its acceptance is much wider in the apricot case. The second case is slightly different from the other two because resistance to mildew and powdery mildew is a prerequisite, and the case shows mainly the work done on the quality of these varieties for the Rosé de Provence. However, along with the third case, it points out that even if the best resistant varieties were created, if they are not adapted to the contexts and do not reach quality expectations, then they would never be grown.

By focusing on varieties in the making, we were able to explore the articulations between phases characterized by a more open-ended perspective and others by a more deterministic perspective. Our case studies dealing with three different processes of definition of ideotypes for plant breeding inform how open-ended and deterministic perspectives are at stake in the case of perennial plant breeding. The creation of ideotypes is intrinsically a deterministic process: objectives are defined with regard to the expected characteristics of the variety. The three case studies have shown how a reflection on future varieties for crop breeding in perennial

---

[17] These may involve farmers through their « representative », rather than « concerned » farmers as was the case here (see below).

Plant breeding for agroecology: A sociological analysis 93

plants can be adapted to the agroecological transition. Ideotypes become varieties through the establishment of socio-technical networks around resistances, qualities, tastes, etc. that address the uncertainties surrounding these productions and their possible ecologisation. These socio-technical networks are dynamic, that is, evolving over time and in different contexts. The long time scale of perennial plant breeding requires frameworks that need to be all the more open that they are intended for agroecological transitions. One reason is that if human collectives around current plant breeding projects are constantly evolving and reconfiguring themselves, so are the main "problems," whether they are expressed in terms of pesticide reduction or adaptation to climate change. However, at some points of the processes, decisions have to be taken (for example, the prioritization of criteria in the third case) which requires more deterministic perspectives. Our three cases show that these plant-breeding processes indeed articulate deterministic and open-ended perspectives.

The first case shows two breeding programs and their position between open-ended and determinist perspectives: the first one on Bouquet varieties had a less deterministic perspective on resistance than the second one on ResDur varieties focused on pyramiding of resistant genes (Fig. 1). But a growing uncertainty about the durability of resistance forces a reopening of perspectives. This is made possible by the implementation of an experimental protocol that puts winegrowers back into the socio-technical network of future varieties.

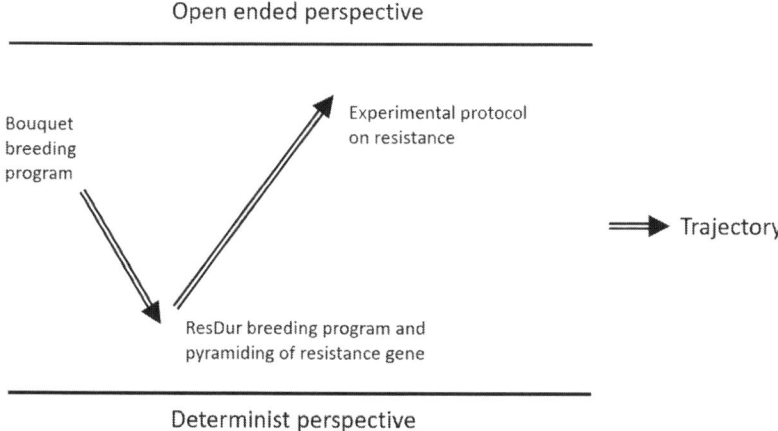

Fig. 1: The trajectory between more open-ended or more deterministic perspectives in the first case study (Languedoc Wine).

The second case deals with a project of varietal creation in Rosé de Provence, resistant to downy mildew and powdery mildew (Fig. 2). While the program of varietal creation is clearly deterministic with the pyramid of resistance genes, the question of taste and how to get closer to the typicity of Provence Rosé wines leads, as in the above case, to re-open the perspectives with regard to varieties.

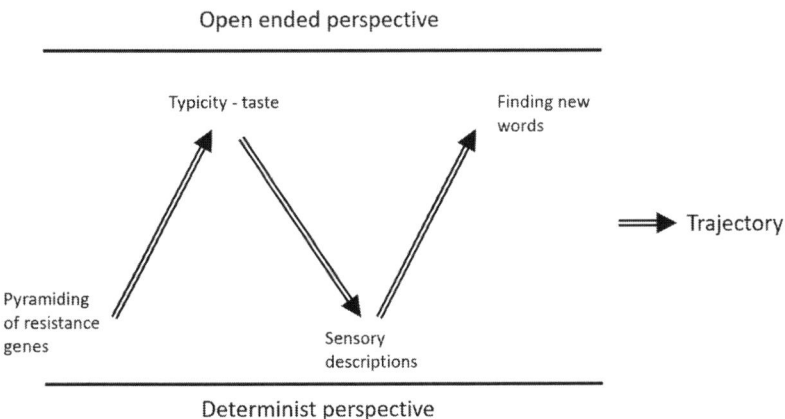

Fig. 2: The trajectory between more open-ended or more deterministic perspectives in the second case study (Rosé de Provence).

In the third case we studied, the actors that face diverse uncertainties (sanitary and economic crisis, market adaptation) are involved in the process of defining the relevant criteria for future varieties and discussed contrasted scenarios able to reflect the diversity of farming and marketing systems and anticipate their possible futures. Through a series of multi-actor workshops, selection criteria were discussed in an open way first, before they were hierarchized based on the discussion of scenarios aimed at addressing the type or degree of agroecological transition at stake and its necessary contextualization, as well as that of the evaluation processes. This translates into a succession of more deterministic and more open-ended perspectives (Fig. 3).

Plant breeding for agroecology: A sociological analysis 95

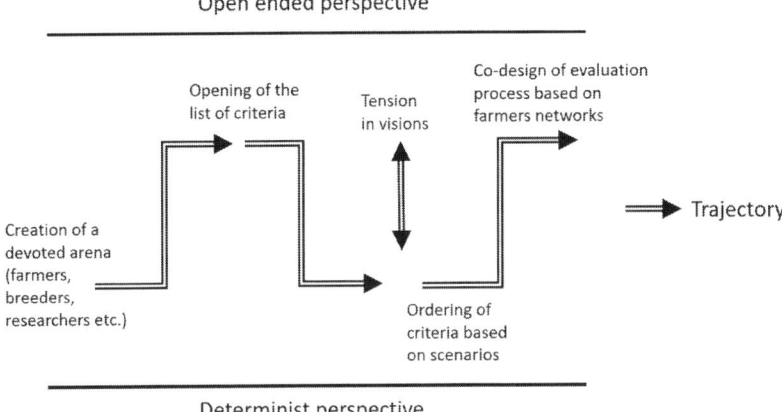

Fig. 3: The trajectory between more open-ended or more deterministic perspectives in the third case study (Prunus).

The three case studies show different forms of articulation or succession between deterministic and open-ended perspectives. The need to design varieties for agroecology forces the actors involved to redefine the problems in both technical and social terms. On the basis of our three case studies and in continuity with the work of Desclaux et al. on the notion of realtype, we have shown the importance of taking into account the socio-technical network in thinking about plant breeding. It avoids the pitfall of considering what belongs to the "social world" and what belongs to the "technical world" in a disjointed manner. Plant-breeding experiments thus appear to be far removed from a vision in which scientists have to mobilize the people concerned to gather needs and information, which they then translate, in their laboratories, into technical problems. What interests us is the tinkering that is carried out to produce the resistance of Bouquet varieties, the typicality of Provence rosé wines, or the innovation system for apricot varieties, where the different actors exchange and jointly reflect on all aspects of the problem. The first case study on the sustainability of resistance to downy mildew and powdery mildew in grapevines shows a constantly evolving socio-technical network. If only the genes are taken into account in a first acceptance of resistance, we observe an opening of the problem to winegrowers who can also act for a more durable resistance. The Rosé de Provence case shows how actors who are not used to working on breeding programs but continuously work on defining what a good Rosé de Provence is, can work with breeders. The socio-technical network of the future variety in the making challenges even the words chosen to describe the wine. Finally, in the apricot case, an important analytical work has been carried out to first share a diagnosis of the past pathways of genetic innovation (related to the trajectory of the fruit industry and of the farms themselves)[18] and then define the perimeter of actors to be included in the collective work and the scenarios to take into account. We observe emerging alliances between the workshops' participants with regard to varieties in the making.

Our wish to discuss the construction of ideotypes for plant breeding in perennial plants for agroecology finally leads us to reflect on the place of the social sciences in these mechanisms. In the apricot project, the social sciences conceived the overall approach with the contribution of the other disciplines – while social sciences are usually contributors rather than "leaders" in genetic innovation projects – and led the various meetings with the stakeholders. In the vine projects, the social sciences were questioned on the adaptation of these varieties for the French wine context and we reformulated a research question around the socio-technical controversies that surround vine breeding. In all three cases, social sciences are mobilized in a context of high uncertainty and in order to "reopen" the deterministic perspective that still dominates in genetic innovation. For instance, the stakeholders involved in the third case study's workshops were chosen because they appeared to be *legitimate* and *involved* actors rather than *representative* ones. They had expressed a strong interest in the debated issues – as was assessed through qualitative interviews prior to their implication – and felt *affected* by the research problem (most had an ongoing collaboration with part of the research team). They also shared a strong *collective attachment* to the future viability of the regional fruit production system (Lamine, 2018). The alternation between more deterministic and more open-ended approaches makes the co-construction of varieties and of their relevant collectives possible. However, we can notice that the socio-technical networks we have studied are not open to consumers and civil society actors. It can be assumed that their integration into these networks would lead to new dynamics and affect the combination of open-ended and deterministic perspectives that we have observed so far.

---

[18] Which led to the co-writing of an article involving researchers and diverse actors involved in the seminars (Lamine et al., 2015).

Acknowledgements

This research was made possible by grants of the Métaprogramme SMACH (INRA), of the Programme Pesticide (Ministry of Ecology) – Prunus project, of the Programme Pesticide/Ecophyto (Ministry in charge of agriculture) – DAS-REVI project, institutional and financial support from SADAPT (INRAE) and Centre de Sociologie de l'Innovation (MinesParistech).

We would like to thank the two anonymous reviewers for their useful feedback. Thanks also to Morgan Jenatton for English-language editing.

References

Bellon, S., Ollivier, G., 2018. Institutionalizing Agroecology in France: Social Circulation Changes the Meaning of an Idea, *Sustainability*, 10, 5, 1380. https://doi.org/10.3390/su10051380.

Belmin, R., Meynard, J.-M., Julhia, L., Casabianca, F., 2018. Sociotechnical Controversies as Warning Signs for Niche Governance, *Agronomy for Sustainable Development*, 38, 44, 1–12. https://doi.org/10.1007/s13593-018-0521-7.

Bessy, C., Chateauraynaud, F., 1993. Les ressorts de l'expertise : épreuves d'authenticité et engagement des corps, *Raisons Pratiques*, 4, 141–164.

Bonneuil, C., Hochereau, F., 2008. Gouverner le "progrès génétique" – Biopolitique et métrologie de la construction d'un standard variétal dans la France agricole d'après-guerre, *Annales. Histoire, Sciences Sociales*, 6, 1305–1340.

Bonneuil, C., Thomas, F., 2009. *Gènes, pouvoirs et profits : Recherche publique et régimes de production des savoirs de Mendel aux OGM*, Editions Quae. Versailles.

Callon, M., 1986. Eléments pour une sociologie de la traduction – la domestication des coquilles saint Jacques et des marins pêcheurs dans la baie de Saint Brieuc, *L'année sociologique*, 36, 169–208.

Compagnone, C., Lamine, C., Dupré, L., 2018. « La production et la circulation des connaissances en agriculture interrogées par l'agro-écologie », *Revue d'anthropologie des connaissances*, 12, 2, 2, 111–138.

Coolsaet, B., 2016. « Towards an Agroecology of Knowledges: Recognition, Cognitive Justice and Farmers' Autonomy in France », *Journal of Rural Studies*, 47, 165–171. https://doi.org/10.1016/j.jrurstud.2016.07.012.

Debaeke, P., Gauffreteau, A., Durel, C.-E., Jeuffroy, M.-H., 2014. Conception d'idéotypes variétaux en réponse aux nouveaux contextes agricoles et environnementaux, *Revue AE&S*, 4, 2, 9.

Delière, L., Schneider, C., Audeguin, L., Le Cunff, L., Cailliatte, R., Prado, E., Onimus, C. Demeaux, I., Guimier, S., Fabre, F., Delmotte, F., 2017. Cépages résistants : la vigne contre-attaque ! Face au mildiou et à l'oïdium, la résistance de la vigne européenne Vitis vinifera ne va pas de soi. Elle est en train de devenir une réalité, au prix d'un travail acharné, *PhytomA*, 708.

Demeulenaere, É., 2014. A Political Ontology of Seeds: The Transformative Frictions of a Farmers' Movement in Europe, *Focaal – Journal of Global and Historical Anthropology*, 69, 45–61. https://doi.org/10.3167/fcl.2014.690104.

Derbez, F., 2018. D'un maïs, l'autre. Enquête sur l'expérimentation collective d'agriculteurs rhône-alpins autour de variétés de maïs population. *Revue d'anthropologie des connaissances*, 12, 2, 259–287. https://doi.org/10.3917/rac.039.0259.

Desclaux, D., Chiffoleau, Y., Nolot, J.-M., 2013. Du concept d'idéotype à celui de realtype: gestion dynamique des Innovations Variétales par une approche transdisciplinaire et partenariale. Exemple du blé dur pour l'AB, *Innovations Agronomiques*, 32, 455–466.

Donald, C.-M., 1968. The Breeding of Crop Ideotypes, *Euphytica*, 17, 385–403.

Gade, D.-W., 2004. Tradition, Territory, and Terroir in French Viniculture: Cassis, France, and Appellation Contrôlée, *Annals of the Association of American Geographers*, 94, 4, 848–867.

Garçon, L., 2015. Réinventer les pommes et les pommes de terre : une géographie de la qualité à l'épreuve des produits ordinaires, *Thèse de géographie*, Université Lyon 2.

Hennion, A., 2005. Pour une pragmatique du goût, Working paper.

Hochereau, F., Clayssens, N., Ugaglia, A., Cristerna-Ragasol, C., Barbier, J.-M., Blonde, P., Touzard, J.-M., 2015. Quel développement des cépages résistants ? *Revue des œnologues*, 157, 28–31.

Lamine, C., 2014. The Complex Relationships Between Breeding Strategies and Sustainable Agriculture: The Tree Fruit Case Study, *International Workshop on System Innovation towards Sustainable Agriculture (SISA2)*, Paris, FRA.

https://hal.inrae.fr/hal-02741618

Lamine, C., 2017. Multi-Actors Transition Arenas in the Fruit Breeding System: A Pathway towards Sustainability, or a New Veil Over Lasting Power Relationships? in Elzen, B., Augustyn, A., Barbier, M., van Mierlo, B. (Eds.), *AgroEcological Transitions: Changesand Breakthroughs in the Making.* http://doi.org/10.18174/407609.

Lamine, C., 2018. Transdisciplinarity in Research about Agrifood Systems Transitions: A Pragmatist Approach to Processes of Attachment, *Sustainability*, 10, 1241. https://doi.org/10.3390/su10041241.

Lamine, C., Mésséan, A., Paratte, R., Hochereau, F., Meynard, J.-M., Ricci, P., 2011. La lutte chimique au cœur de la construction du système agri-alimentaire, in Ricci, P., Bui, S., Lamine, C. (Eds.), *Repenser la protection des cultures*, QUAE, Versailles.

Lamine, C., Pluvinage, J., Aubenas, R., Faugier, V., Simon, S., Clauzel, G., Lamberet, M., Penvern, S., Stévenin, S., Buléon, S., Garçon, L., Bui, S., Audergeon, J.-M., 2015. Innovation variétale en Prunus, 1960–2013 : les enseignements d'une analyse socio-historique co-construite avec les acteurs, Le Courrier de l'environnement de l'Inra.

Litrico, I., Bonnin, I., Duc, G., Enjalbert, J., Goldringer, I., Rolland, B., Ronfort, J., 2014. Agroécologie, Génétique végétale et Amélioration des plantes, Rapport d'analyse Département BAP, Inra. 41p. ⟨hal-01263558⟩.

Méadel, C., Rabeharisoa, V., 2013. Chapitre 7. Le goût comme forme d'ajustement entre les aliments et les consommateurs, in Callon, M., Akrich, M., Dubuisson-Quellier, S., Grandclément, C., Hennion, A., Latour, B., Mallard, A., Muniesa, F. (Eds.), Sociologie des agencements marchands : Textes choisis, Sciences sociales, Paris, Presses des Mines, 171–194,. http://books.openedition.org/pressesmines/2030.

Sylvander, B., Bellon, S., Benoit, M., 2006. Facing the Organic Reality: The Diversity of Development Models and Their Consequences on Research Policies, *Paper Presented at Joint Organic Congress*, Odense, Denmark, May 30–31, 2006.

Teil, G., 2009. A la recherche du plaisir : Une analyse pragmatique de la perception, in de la, M. S. H. (Ed.), *Ethnologie des gens heureux*, Cahiers d'ethnologie de la France, 210 p.

Vanloqueren, G., Baret, P., 2004. Les pommiers transgéniques résistants à la tavelure analyse systémique d'une plante transgénique de « seconde génération », *Courrier de l'environnement de l'INRA*, 52, 5–21.

Agroecological transitions at the scale of territorial agri-food systems

Marianne Hubeau, Martina Tuscano,
Fabienne Barataud, Patrizia Pugliese

1. Introduction

Diverse challenges such as climate change and the need to retain consumer trust highlight the necessity for agroecological transitions. Changes in institutional settings and practices are required to respond to these challenges. Within this chapter, we focus on agroecological transitions at the scale of territorial agri-food projects. We apply a holistic approach which allows a multilevel, multiscale and multi-actor approach taking the dynamics of the territorial agri-food system and its interdependencies with the natural, human and environmental system into account (Haberl et al., 2009; Binder et al., 2010; Lamine, 2011). The studied projects are implemented at the territorial level, understood here as a scale for thought and formalizing action (Kloppenburg et al., 1996). The territorial scale as well the local scale must be understood as a "social construct," which produces strategies pursued by actors by putting a public problem on the agenda (Born and Purcell, 2006; Garçon et al., 2017). We adopt a territorial approach to study agroecological transition projects. This moves beyond a globalized model while aiming to reinforce the capacity of agri-food projects that integrate territorial resources such as land and water but also landscape conservation, education, and social relations of proximity (Renting et al., 2003; Lamine et al., 2019). Furthermore, we aim to uncover processes, relationships and governance structures (Sonnino et al., 2016; Stotten et al., 2017). These new governance mechanisms may pool different resources such as space, place, people or expertise to find innovative solutions. Additionally, in practice, the territorial actors, such as local authorities, producers, consumers and citizens, use place-based knowledge and are acting upon broader agroecological issues such as food and environmental security, safety and sustainability.

In this chapter, we studied governance through territorial network structure, dynamics, actors, processes and interactions, focussing on open-ended (OE) versus deterministic (D) perspectives. Our main objective is to study how agri-food projects aiming to foster agroecological transitions focus on a specific configuration of OE/D perspectives and how this is adapted to a specific territorial context. In order to explore to what extent these agroecological transition projects take the different elements of their territorial context into account, we analyse four multi-actor, territorially diverse projects. The territorial approach is relevant both as a context in which action is performed and is oriented, as well as a methodological device to identify, empirically, the actors involved. It is useful as a category of action and as a category of analysis. The need to include new actors and projects in food-related issues and the complexity of agroecological transitions emphasizes the importance of new food governance systems (Rossi, 2017). The OE/D perspective allows addressing this issue by focussing on the perspectives of the actors involved within these projects. In order to assess OE/D perspectives, we develop a conceptual framework to analyse their tensions. Specifically, we analyse the governance types at stake through how objectives are defined, how projects are governed and how processes of consideration and/or exclusion of actors are managed. In that sense, OE/D perspectives are categorized in OE/D objectives and OE/D processes. The reason is twofold: first, scientifically, these projects aim to break with existing long-standing governance mechanisms and therefore we aim to understand this transition process. Second, politically, we aim to understand how the goal of an agroecological transition is applied and is adopted within a specific territorial context. The second section describes the conceptual framework and section three the trajectories of four agri-food projects in four regions, namely Flanders (Belgium), Provence Verte (France), Mirecourt (France) and Parco delle Dune Costiere (Italy). Finally, Section 4 discusses some tensions and controversies identified within these territorial agri-food systems.

2. Conceptual framework

The conceptual framework allows us to analyse our four territorial trajectories based on how they are conceived by their protagonists and through the lens of the tensions between OE/D perspectives regarding agroecological transitions. Because we analyse four agri-food projects whose aim is to transform the way in which actors produce, process, supply and consume food, the notion of "project" requires elucidation. Recent studies stressed the need to take the notion of "project" into account in alternative food networks (Velly, 2019). Often, initiatives dealing with the greening of the agri-food system are categorized as "alternatives," casting aside the heterogeneity of demands and forms of cooperation in contrast to the dominant agri-food system. Moreover, this type of research often ignores initiatives that combine "conventional" and "alternative" patterns (Lamine et al., 2019). We consider projects as an analytical category that may reflect hybrid dynamics. Hence, we look at these projects as new governance mechanisms including dynamic processes reflecting the performative power of the objectives and steps defined by their protagonists. In addition, project analysis makes it possible to explore the capacity for collective organization and therefore to see how governance modes concretely permit to perform agroecological transitions at the scale of territorial agri-food projects.

We analyse these projects from a territorial perspective including a conceptual as well as methodological viewpoint. Territory consists of a given reality in which stakeholders seek to drive transitions. Projects allow creating links through a normative activity between the actors and the physical and socio-economical components of the territory. Moreover, the territorial approach makes it easier to identify the actors of the agri-food system and to analyse the existing interactions between its links, that is, the actors of the agricultural world, public policies, economic actors, civil society and research (Lamine, 2012; Bui, 2015). Our conceptual framework is illustrated in Fig. 1.

We focus on the evolution of governance and multi-actor processes in relation to transition perspectives. Indeed, our assumption is that OE/D perspectives result from the combination of visions and arrangements in a given context. The concept of governance is often used in research about agri-food systems to study their organization, coordination and functioning. Governance could be conceptualized as the structure including the involved actors, their role, the formal and informal

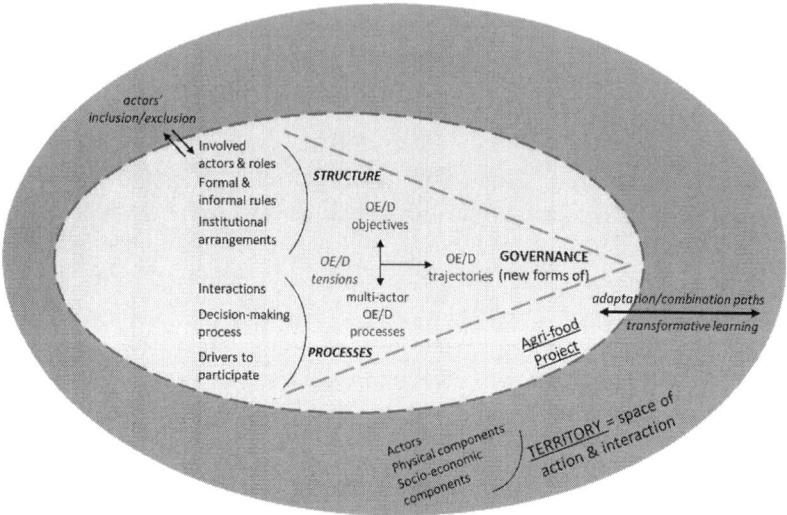

Fig. 1: Interplay of territory and agri-food project(s) in agroecological transitions.

rules, institutional arrangements, and processes such as interactions and decision-making processes (Folke et al., 2005). As some authors have underlined, transformations in agri-food systems demand a reflexive governance approach for research, society, policy and industry, while integrating a diversity of perspectives, discourses, visions and goals (Marsden, 2000, 2013). Key elements in such an approach are participation, collaboration and collective learning. Within territorial agri-food projects, multiple actors collaborate within the same spatial scale and societal context. This multi-actor collaboration could enhance learning, co-produce knowledge and increase transformative capacity to address complex sustainability problems and explore competitive advantages (Borgatti and Molina, 2003). The organization of multi-actor processes with a diversity of heterogeneous visions is challenging for all involved actors, such as researchers and practitioners. Although the complete steering or coordination is impossible, previous research suggests that the speed and direction towards new practices such as agroecology can be directed (Loorbach et al., 2017). We will discuss to what extent these multi-actor processes and transformative learning can enhance agroecological transitions by looking at the four agri-food projects through the lens of OE/D perspectives. This iterative process should organize the process, structure the governance mechanisms and its dynamics without directly controlling it (Turnheim et al., 2015).

For analytical purposes, we have divided OE/D perspectives into OE/D objectives and OE/D processes. The former correspond to shared visions translated into project actions while the latter correspond to ways in which actors organize and coordinate themselves. D objectives correspond to cases where precise targets and goals are defined while OE objectives have no precise targets and goals. OE processes are open to all interested actors and actively try to involve a diversity of actors while also redefining the governance process including collective decision making processes. A D process is more predetermined in terms of involved actors, governance structure and process. This analytical framework does not aim to demonstrate that one approach would be more or less appropriate than the other, but rather to discuss the issues at stake in these two types of processes as well as the existing tensions between visions, objectives and the consideration of stakeholders. Through this framework, we identify and qualify the conditions under which OE/D tensions crystallise and the conditions that allow adjustments in projects in terms of visions, involvement and exclusion of actors, or governance modes.

3. Method and case study description

All four projects focus on agroecological transitions in different ways; these differences formed one of the main criteria used to choose them. The lack of uniformity in trajectories, spatial scale and processes make it interesting to explore how transition mechanisms can emerge and are carried out in different contexts. Further, these case studies were chosen based on the geographical heterogeneity in which they have developed as well as the diversity of socio-economic contexts. All case studies are longitudinal case studies (Yin, 2003). Each author of this paper conducted an in-depth study of one of the four case studies. Although methods in fieldwork differed slightly, in every case, qualitative methods were used. Tab. 1 describes the data collection methods of every case study. In all cases, the involved researcher could draw on long-established contacts and previous research.

We describe the four projects based on their trajectories including the history, involved actors, objectives, and governance modes. Through these trajectory narratives, our aim is to show, in a descriptive and

Tab. 1: Data collection in four agri-food projects.

Agri-food project region	Data collection
Flanders	Document analysis: website, media articles, policy documents; Eight semi-structured interviews: Input supplier (1); Food industry association (2), Producer (1); Distributor (1); Policy actor (1); NGO (2); Participant observation: Discussion groups (15); Workshops (5); Focus groups (9)
Provence Verte	Document analysis: four agricultural and food policy documents; Participant observation of discussion groups (8), workshops (4) Four semi-structured interviews: food project moderator (1), food project policy actor (1), civil society associations (2)
Mirecourt	Document analysis: minutes of meetings and statutes of associations Participant observation of discussion groups (20) Four semi-structured interviews: organic farmers (4)
Coastal Dunes Park	Document analysis: policy documents, website, flyers, brochures media articles; Participant observation; 4 semi-structured interviews: Park Director (1), Tourist operator (cooperative offering tourist services) (2), producer (1), restaurant (1)

inductive way, how these OE/D objectives and processes are concretely translated. In doing so, our core objective is to contextualize the territorial agri-food projects to avoid adopting a normative stance, and instead analyse in an inductive way whether the perspectives adopted by their protagonists are either OE or D. We also attach particular importance to turning points during the project development as favourable moments for adjustments in coordination and transition vision. A turning point is a specific moment initiating a change in governance mode, the actors involved, and indirectly a change in OE/D perspectives. The narratives allow us to understand how the strengthening of interdependencies as well as controversies between actors in agri food systems over time can lead to these transition processes. Each narrative starts at the beginning of the project and follows a similar structure with the aim of relating the objectives, the actors involved and the formal structure of the project. Then, through a dynamic approach, we look at changes and adjustments over time.

3.1. Step-by-step evolution towards a deterministic perspective in Flanders

The timeline of the food system trajectory in Flanders is represented in Fig. 2 which illustrates three agri-food projects. Flanders is the northern part of Belgium, with a high urbanization rate and a high population density. Most farming systems are specialized (88% of the farms) and intensified with indoor production systems. On a territorial level, animal feed takes the biggest share (56%) followed by arable farming including grains, potatoes and sugar beet (34%).

New food frontier started in 2010, when two academics, two NGO representatives and a policy maker started a transition process aiming at convincing a large group of actors. Hence, a long-term vision and an action framework were essential. Moreover, the process involved both creative thinkers as well as practitioners (Fig. 3). The initiators aimed to create an open governance structure by setting up a policy-supported network. The steering group consisted of local and national administrators, NGOs, a researcher and two supply chain actors; they agreed that they would only communicate in case of full consensus. In 2012, five deliberative sessions were organized to formulate future "images" of the Flemish agri-food system resulting in three images that clearly represent distinguished discourses, namely ecological modernization, de-commodification emphasizing consumer-producer relations and a sufficiency discourse focussing on new cultural relations. A first turning point occurred in 2012, when after the third deliberative session, the supply chain actors decided to leave the initiative with a motion of distrust considering all outcomes as illegitimate. The open-ended objectives and process caused tensions between the various actors because their objectives were dissimilar. For instance, an article of three members of the steering group appeared in the agricultural press that described ongoing production processes as a "failure"of the system. The supply chain actors disagreed with the description of the system as being-in-error and the agreement of communication in full consensus was broken; this conflict ended the *New Food Frontier*.

Subsequently, the supply chain actors recognized the urge for a sustainable system and started a new initiative in 2013, the *Transformation project*[1]. This project had deterministic objectives, namely to identify

---

[1] *Het Transformatieproject*; for an in-depth analysis see Hubeau et al., 2017, 2018.

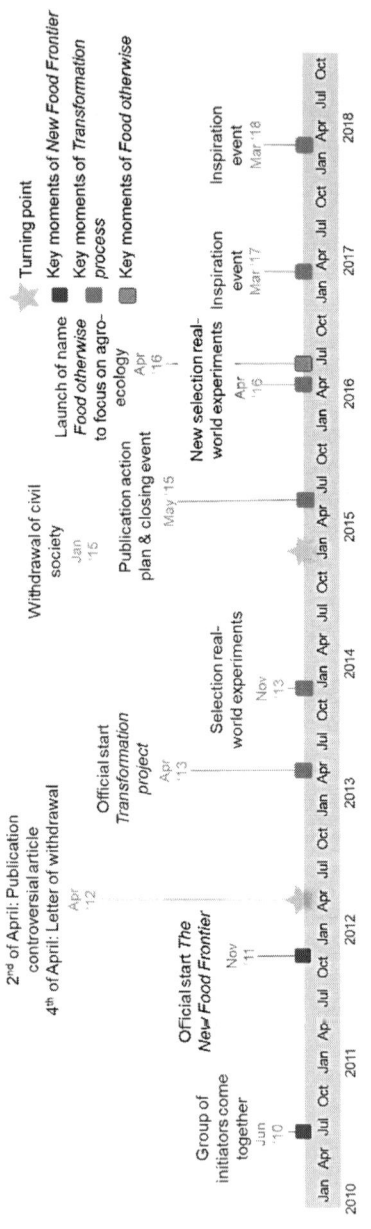

Fig. 2: Food system trajectory in Flanders: key moments and turning points.

shared transformation pathways and to develop an action plan. The main differences with the *New Food Frontier* were (1) the initiators are the supply chain actors who took an active role managing the process, (2) the explicit focus on transition changed to transformation defined as a stepwise evolutionary process inspired by niches, and (3) the initiators emphasized real-world experiments instead of shared vision development. The initiators asked researchers to govern a transdisciplinary approach. The three actors of *New Food Frontier* responsible for the media article continued to participate. Nevertheless, they took another role. The NGO representative was part of the steering group, the researcher was part of a scientific committee and the policymaker was part of a broad stakeholder group. Although the objectives were deterministic, the process was still open and attempted to involve a broad range of actors such as industry, intermediaries, research, civil society and policy (Fig. 3). In addition to meetings of the steering group occurring at least every two months, multiple workshops, focus groups and discussion groups were organized, on average every four months.

However, the combination of precise targets and an open process caused conflicts and a second turning point occurred in 2015. At that time, the NGOs grouped in a separate network. Just before the release of the action plan, they collectively distanced themselves from the *Transformation project*. Despite a search for solutions, no reconciliation was possible as the differences in objectives of the NGO network and the *Transformation project* were insurmountable. For instance, a conflict arose about the position of meat production and consumption in the Flemish agri-food system. Strategic purposes arose as some NGOs stated that further participation in the initiative would decrease their legitimacy and could reduce their credibility as social movement campaigning for radical agroecological transitions.

After this second turning point, two parallel processes further developed. The *Transformation project* continued while the NGO network, under the name *Food Otherwise*[2], continued with a full emphasis on agroecological transitions. The main difference with the previous phase is that both processes set deterministic objectives and partly closed the process by excluding each other in their parallel processes (Fig. 3). For instance, *Food Otherwise* aimed to reach one specific type of

---

[2] *Voedsel Anders.*

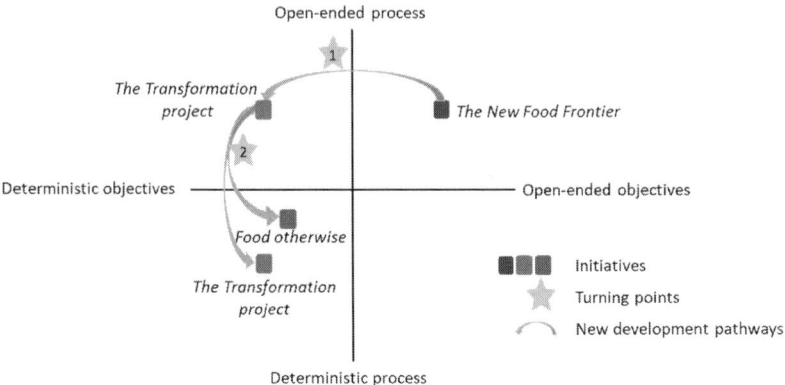

Fig. 3: Trajectory of Flemish case: step-by-step evolution towards deterministic objectives and process.

agroecological practices, namely labour-intensive agricultural practices as opposed to the three discourses targeted in the *New Food Frontier*.

3.2. Provence Verte food project: From open-ended to deterministic perspectives

Provence Verte is a group of 28 municipalities situated in a rural area in the southeast of France with approximately 100,000 inhabitants. Although urbanization has been increasing since the 1970s, agriculture still is an important economic sector of the region. The main agricultural activity is wine grape production occupying about half of the agricultural land. In 2017, organic production represented 22.9% of farmland and the process of conversion to organic is constantly increasing. The territory benefits from multiple initiatives carried out by local authorities[3], by civil society, and by supply chain actors fostering alternatives production and consumption practices. Nevertheless, until very recently, no territory-wide strategy linked nor supported initiatives steering agroecological transitions.

---

[3] For example, the case of the village of Correns where almost all agricultural land is organic.

Territorial agri-food projects' trajectories

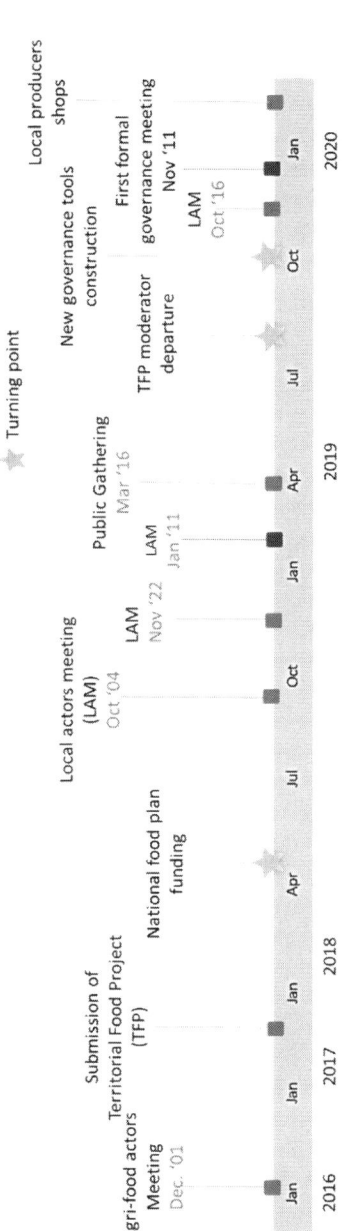

Fig. 4: Timeline of Provence Verte trajectory and its turning points.

The Provence Verte food project started in 2016 (Fig. 4) when the *Provence Verte Development council* organized an agricultural meeting inviting numerous actors engaged in agroecological practices. At this event, the guidelines of the Territorial Food Project (TFP)[4] were outlined. The project aimed to respond to the "desire to restore the economic and social life of an agricultural sector in transition by diversifying and sustaining local agricultural activities, promoting employment and gradually building a TFP that would meet the needs of the territory." Initially it was also conceived as a collective co-construction dynamic based on "inclusive and participatory local governance."

In May 2018, the project received funding from the National Food Plan (NFP) in order to engage a moderator and finance project actions (Fig. 5). The TFP main drivers are a recently created local authority – which agglomerates the 29 municipalities[5] – and the Agricultural High School of St Maximin, a central player engaged in agroecology practices by educating students. For a formal necessity requested by the call for proposals, the project's initiators had to define operational components, list the actors to be involved in the process and give objectives to be achieved based on the funding. Even if the main objectives and involved actors were defined in the TFP, the desire was clearly to undertake an exploratory and participative process to allow agri-food transition in the territory. Therefore, objectives, as well as steps to follow, should be built along the way, in an open-ended perspective.

Actors involved in the early meetings of the TFP were mainly the "alternative" actors of the region, namely the organic farmer's association, environmental associations, support structures for farmers and some producers and retailers. The need emerged for existing operating actors to connect the several alternative stakeholders and initiatives steering agroecological transitions in the territory. Therefore, the funding granted to the TFP could be used for this purpose. When looking through the lens of OE/D and governance, the question of how to make these stakeholders collaborate is immediately identified as a crucial issue. The moderator and key stakeholders proposed an innovative procedure to work together via a public gathering. They made the hypothesis that a "natural food

---

[4] Territorial Food Projects are introduced by agricultural law in 2014 with as main goal creating a local food strategy.

[5] Communauté d'agglomération de Provence Verte.

governance" could be enhanced within the territory by mobilizing local stakeholders around the organization of this public event. The *modus operandi* chosen was therefore very exploratory and *a priori* excluded any pre-established form of governance (e.g. a steering committee). The event took place in March 2019 and only few inhabitants and actors involved in agri-food system transition took part. The approach chosen created dissatisfaction among the local actors who took part in the project as they did not see concrete goals in the chosen process.

The moderator decided to abandon the coordination of the project. We consider this a turning point in the trajectory of the project (2). The main stakeholders decided to change radically the project's direction, clearly defining the objectives to be achieved and creating an official steering committee. This shift in direction has benefitted from the creation of a Local Development Council with an attached "agriculture commission." This group of citizens wished to be involved in the TFP, as it was the main territorial agri-food project held in the territory. During the months of October and November 2019, the Agricultural High School of St Maximin begun to work with this citizens' group in order to identify actions to be carried out within the framework of the TFP. During the first official steering committee in late November 2019, the Agricultural High School presented the work carried out together to the Agriculture Commission, to many regional institutional and associative stakeholders as well as to some organic producers. The operational decisions made by the project leaders related to access to land, support for the installation of young farmers and the relocation of farm-based catering. The TFP of Provence Verte has therefore evolved from a very open and exploratory process of governance and objective definition to a rather deterministic set of objectives and guidelines defined by a circumscribed group of actors.

3.3. Mirecourt: A test of a user-centred approach that emphasizes action and values

The Mirecourt-Dompaire urban community consists of more than 70 small municipalities located in western Vosges, France. Mirecourt is the main municipality of this community with 5,000 inhabitants. Industry is in decline and depopulation is ongoing. The area is classified as a "rural revitalization zone." The agriculture is polyculture-livestock which remains a stable sector but requires a specialization at farm level

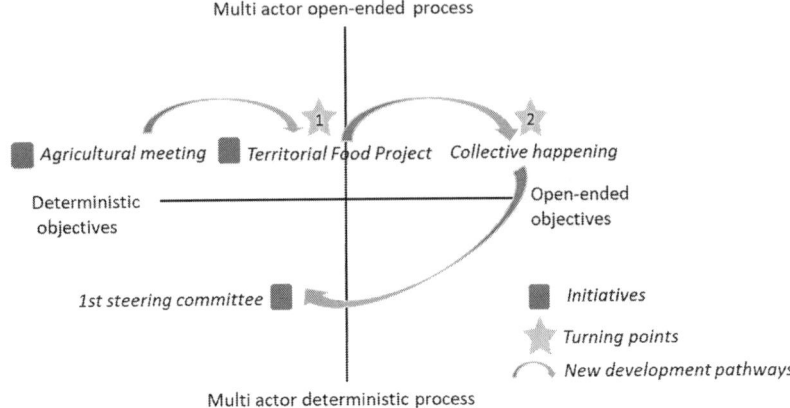

Fig. 5: Trajectory of Provence verte: from open-ended to deterministic perspectives.

(with suckler cows on grassland and dairy cows fed with maize and cereals). The major products valued in short supply chains come from a "minority" of diversified production systems such as sheep and other herbivores, beekeeping, etc., which do not benefit from regional collection or processing facilities.

The trajectory is based on the convergence of three dynamics driven by different actors. First, the "Bios" producers' group was started in 2013 by a group of eight organic pioneer farmers who actively resisted forms of "conventionalization" or "intensification" through short market channels by reconnecting consumers with food producers. A second initiative carried out by INRAE[6] is based on its experimental organic farm of 240 ha whose aim is to find complementarities across its productions and to build strong networks with local producers (Coquil et al., 2019). Finally, the third initiative, a "citizen café" in Mirecourt, l'Utopic, emerged in 2015. The founding members organize debate evenings within which the theme of agriculture and food arose. These three initiatives were linked through personal relations or professional and associative commitments. These initiatives converged to create a project submitted to the Fondation

---

[6] Research unit "ASTER" of INRAE (French national research institute for agriculture, food, and environment).

of France (FdF)[7] in 2015 for supporting the development of a local and sustainable agri-food system in Mirecourt.

The project's aim was to promote access to healthy and sustainable food and the creation of income-generative agricultural activities by non-relocatable jobs. The INRAE unit immediately played a dual role as agricultural producer and as a reflexive actor formalizing governance and project management. INRAE proposed to adopt a living-lab approach, that is, a user-centred approach for societal innovations. A project manager recruited with FdF funding (hired by INRAE) was in charge of the coordination of the process. An open and horizontal governance model was sought with some shared guidelines such as working around shared values before defining actions while including a diversity of actors. This inductive action based on opportunities of initiatives is inclusive and horizontal, framed by values rather than operational objectives. In this sense, it can be qualified as OE more than D. On the other hand, the point of view shared by the collective that organic farming must be a privileged form of agriculture to achieve the major objectives provides fairly a strong framework for initiatives (and could be qualified as a form of D).

New actors were soon included in the project, for example, the farm of the local Agricultural High School, the local group "Rural Households" (an association of popular education in rural areas[8]), and an NGO[9] aiming to stimulate social integration of non-francophone people. Extending participation to new actors was encouraged by multiple factors: the presence of people belonging to different worlds, the conference cycle at the citizen café, the territorial insertion of INRA through its experimental farm, the project management mode. The flexibility of governance also favoured these extensions. No precise target nor a decision-making committee had been pre-established. One significant conflict arose when a key actor in the citizen café quickly adopted operating methods that offended a large part of the collective and caused severe tensions. These tensions crystallised around requests for funding made by this actor involving all the partners without them being informed. The two actions planned without consultation were a citizen café festival (planned and

---

[7] Fondation de France.
[8] Fédération des Foyers Ruraux des Vosges.
[9] La Vie Ensemble.

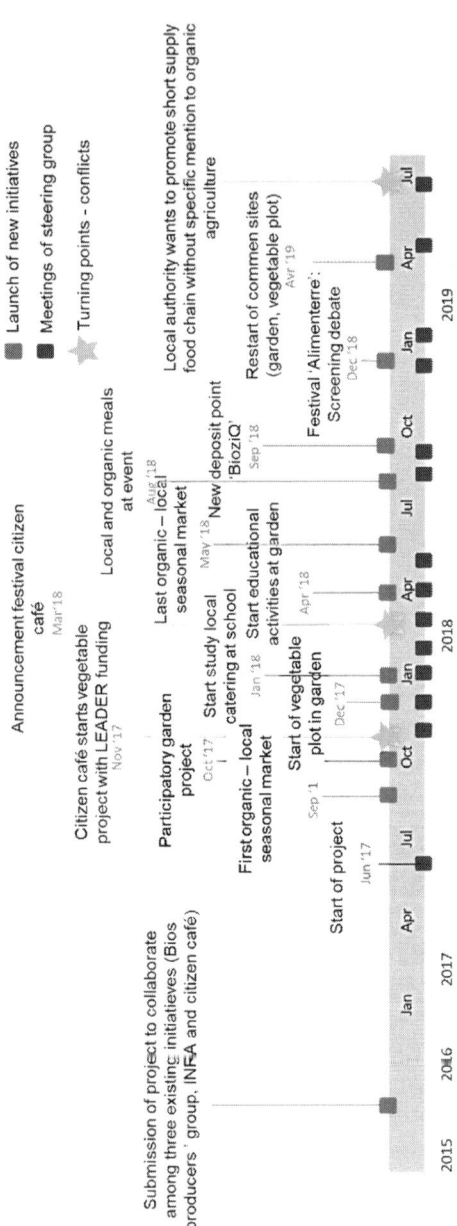

Fig. 6: Timeline of the Mirecourt trajectory.

Territorial agri-food projects' trajectories

carried out in his name) and a processing unit project (the collective doubted its feasibility – it never came to fruition despite having received funding).

From 2016 to 2019, several actions were implemented in practice. Some actions were directly linked to one of the initial aims (organic producers' depots) according to a rather D process, while other actions arose in the process with an OE character. For instance, a plot of open field vegetables at the INRAE experimental farm was designed and collectively cultivated by a collective of INRAE staff, volunteers, clients of three NGOs (the social integration NGO mentioned above and two food aid associations[10] widened the circle of partners), as well as people with mental disabilities who were clients of a medical-social work organization[11]. A side aim of these partners was to demonstrate the feasibility of such an operation and to engage the municipality in a similar approach for the collective restoration of the city's school buildings. The district planning authority thus recently joined the dynamic by taking the official management of a TFP label application. This new political shift could create new tensions in light of the diversity in understandings of project management and agricultural orientation.

3.4. Spiralling up and out: The ECST experience in Coastal Dunes Regional Nature Park, Italy

Established in 2006 the Coastal Dunes Regional Nature Park spans its territory over two municipalities in Puglia, covering an area of 1,100 hectares, popular for its coastline and inland agricultural landscape speckled with centuries-old olive groves, historical farmhouses and archaeological sites where traditional extensive farming co-exists with patches of CAP-induced intensive agriculture. Fig. 8 shows a cross-sectional snapshot of the Park's experience from 2012 to 2018.

In 2012, the Park was first awarded by the European Charter for Sustainable Tourism (ECST) certification for Protected Areas[12]. Recognition as an ECST *sustainable destination* was successfully renewed in

---

[10] Restos du Cœur and Popular Relief.
[11] Structure for integration through work, whose branch in Mirecourt is specialized in market gardening activity.
[12] https://www.europarc.org/sustainable-tourism/

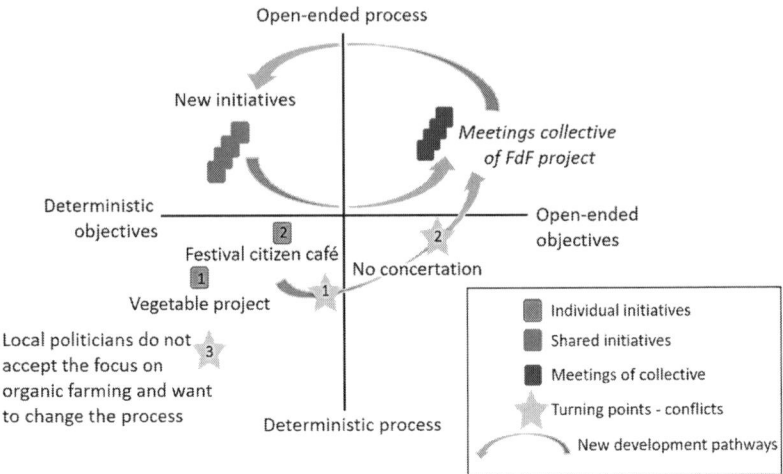

Fig. 7: Trajectory of Mirecourt: testing a user-centred approach that emphasizes action and values.

2018, while in 2015, 21 local businesses operating within the Charter area had already been awarded the ECTS certification as *sustainable business partners*. Coastal Dunes was one of only two Italian parks to first complete the two parts of the Charter program.

The ECTS recognition certifies a voluntary, multi-actor, stepwise process, participatory involvement of the protected area, its community and the tourist value chain. Inspired by the 5 Charter principles, it addresses 10 key topics that are translated into concrete actions within a shared strategic vision and a co-designed five-year action plan that is internally monitored and externally evaluated.

The Charter's call for *'quality sustainable tourism (…) good for Parks and good for People'* matched well with the Park director's vision. The ECTS programme provided an opportunity to engage into a structured path to reframe local stakeholders' commitment, embed and coordinate individual/collective engagement for an integrated sustainable development.

In parallel, the Charter program's flexibility allowed the Park to extend beyond a narrowly determinist, linear approach. It left room for circularity and iterative exchanges within the original core group of stakeholders and a gradually widening circle of actors which brought in in

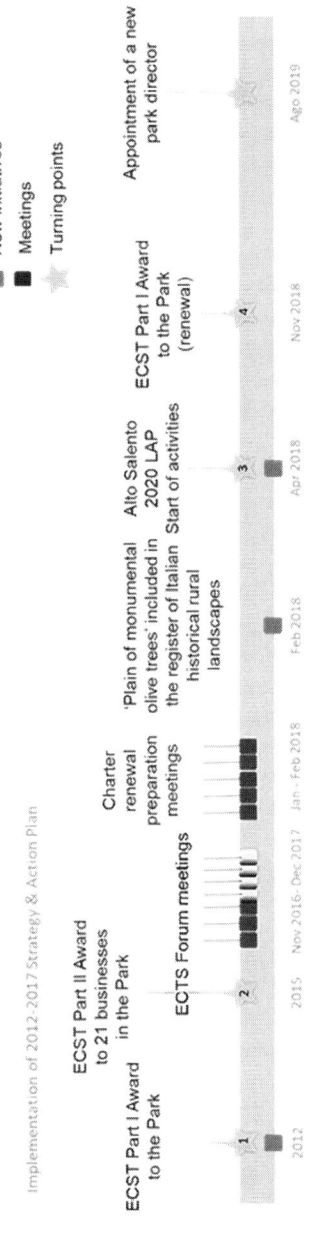

Fig. 8: Timeline of the trajectory of Coastal Dunes Regional Nature Park.

new perspectives. Two hundred stakeholders participated in ECTS plenary forum meetings for co-designing 2018–2022 Strategy and Action Plan and over time, the number of operators that regularly collaborated with the Park to implement identified actions doubled: an evidence of a positive open-ended *spiralling up* dynamic (Fig. 9).

The Charter's framework represented a transition arena favouring relational reflexivity and offering opportunities for reflexive governance. Continued interaction helped to build trust, consolidate existing and establish new working relations and learning networks, recombine discourses and belongings, experiment with innovative business ideas, exercise co-responsibility for protection and management of local human, social and environmental capital. Progressively, a new model of territorial governance emerged, revolving around a pro-active role of the Park authority that was no longer perceived as a normative institution (in contrast to the years prior to 2012). A tangible sign of enhanced Park-operators partnership was the uncontroversial adoption of revised guidelines for granting and monitoring the use of the Park's label. This was awarded based on contractual arrangements to farms, restaurants, tourist operators complying with quality and sustainability criteria or certifications. The Park is a laboratory of good practices catalyzing open-ended dynamics of innovation. Application of the revised guidelines will

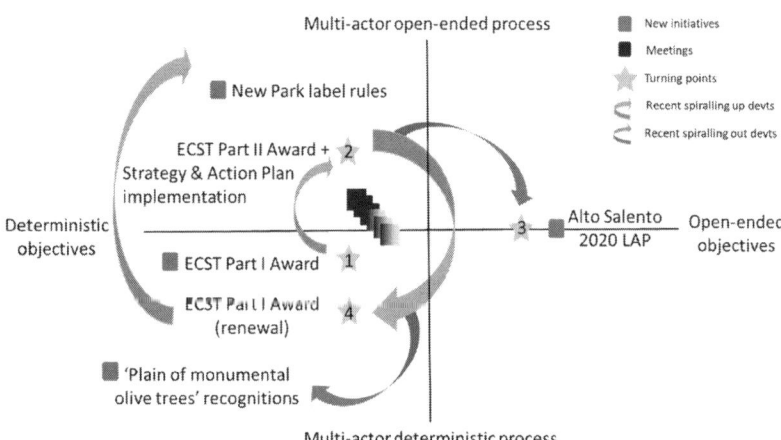

Fig. 9: Trajectory of the Italian case: engagement for local development spiralling up and out.

Territorial agri-food projects' trajectories

further reshape practices: diffusion of organic farming and agroecology, promotion of quality foods (organics, Slow Food presidia, PDOs) and green tourist services reconnecting "the rural" and "the coast," integrating food territorial identity with local culture, art, slow mobility and environment protection. Besides the main spiralling up path, Fig. 9 also shows positive *spiralling out* aspects that articulate the Charter approach and outcomes with other territorial frameworks through inter-scalar connections that led to the scaling up of ECST experience to a geographical area extending beyond the Park's boundaries.

In the framework of the EU Leader Plus Initiative, Alto Salento Local Action Group started its activities in 2018 with a focus on one central topic: rural and coastal landscape care and management to support responsible and slow tourism. Much of ECST experience clearly underpins the Salento 2020 Local Action Plan. The appointment of the Park's director as LAG director favoured synergy between the two initiatives. The Park also played a pivotal role in the application for the inclusion, in the Italian National Register of Historical Rural Landscapes (recognition awarded in 2018), of the centuries-old monumental olive grove landscape in Puglia (including the Park's area and two other neighbouring municipalities), and also in the preparatory work for the application for recognition of the same area as GIHAS (Global Important Heritage Agricultural Systems).

A new Park director was recently appointed. If the Park's achievements had been strongly connected to the former director's charismatic personality and his professional experience as rural development facilitator, stakeholders' collaboration with the new leadership and continuity in vision and action will be necessary to pursue constructive institutional dialogue and effective public-private partnerships.

4. Results and discussion

We have studied four agri-food projects to analyse their specific configurations of OE/D perspectives in relation to their specific contexts. The OE/D perspectives were studied on the basis of the OE/D objectives and OE/D processes as we analyzed how objectives are defined, how the projects are governed and how processes of inclusion and exclusion of actors are managed. We primarily discuss empirical results of our case comparisons. Subsequently we provide insights for future research and exploitation of lessons learnt from a practitioners' angle.

First, we observe turning points in all four cases. In three of them (Mirecourt being an exception), a turning point initiated a change in governance mechanisms. Usually, this turning point was directly linked to a conflict that created tensions between actors within a network. In the studied trajectories, conflicts were mostly related to differing objectives regarding the direction of the transition (e.g. Flanders), a mismatch between expectations and the complexity of reality (e.g. Provence Verte), or a lack of transparency (e.g. Mirecourt). A turning point also leads to more reflexivity regarding objectives and processes. In three of the agri-food projects (except Mirecourt), the projects reactively adapted their objectives after a turning point instead of proactively adapting based on reflexivity. This result suggest that proactive reflexive practices are far from obvious in transition processes, as mostly reactive reflexivity occurs. The Coastal Dunes Park case was more proactive due to the ECST cycle which provided structure and ensured continuity for a long-term integrated local development vision embedded in previous initiatives. Our results confirm other studies (Marsden, 2000; Hubeau, 2019) that highlight the importance of reflexive practices within the context of transitions. More specifically, the core idea is that there should be enough self-reflexivity within the process for the actors to remain open to multiple perspectives, to initiate continuous questioning and to put an emphasis on interactive processes that mobilize the knowledge and resources (Binder et al., 2015). Moreover, such a reflexive approach must enable the collaborating actors to frame and tackle persistent problems. Some specific strategies that may be applied are transdisciplinary knowledge production, real-world experimentation and iterative participatory vision forming (Voß and Kemp 2006; Smith and Stirling 2008) which is also observed in our trajectories. For instance, in Flanders, a transdisciplinary process was set up to develop a strategic action plan; in the Coastal Dune Park, the ECST process has provided the opportunity to draft and implement a strategy and an action plan in a participatory and inclusive manner; and in Mirecourt, multiple initiatives were initiated as experiments.

Second, we observe a significant role of personal relationships within the various projects. The personal bonds could either accelerate or delay the transition. Our comparison shows that a highly exploratory and therefore OE approach may not be appropriate in a context without an existing network of operational actors. Indeed, it can be difficult to get involved in a process with actors you know little or nothing about without seeing the direction and methodology of the process. This could delay the active development of a shared culture at the initial phase. Other researchers (Claro et al., 2003) also highlight the importance of trust and personal bonds within multi-actor processes. Commitment to joint/shared action can be a good way of building this better mutual knowledge and trust.

Third, the personal characteristics of certain actors, their role or the role of certain organizations are an important element. Mirecourt is an illustration of this, as INRAE plays an important role as a bridging organization throughout the whole project because of its multiple roles, its knowledge of the network of actors, and its role in stimulating reflection among actors. This confirms other authors' description of the role of bridging and boundary organizations as anchoring points that can accelerate transitions (Folke et al., 2005; Klerkx and Leeuwis, 2009).

Fourth, OE/D perspectives and processes also play a different role during different project phases. We observe some trends of OE/D perspectives throughout the different project phases. For instance, during project writing or exploration, only a few actors are involved (e.g. in Provence Verte no actors were involved). Furthermore, it seems important that at the beginning of the project the actors involved are encouraged to state which transition process they wish to generate. More specifically, they should agree on the objectives they want to achieve and the transformation processes undertaken, as these are interlinked. In other words, a shared culture including shared values, shared practices and shared narratives is necessary within agri-food transition projects (Hubeau, 2019). A shared culture helps to create an internalized decision-making process and a transparent communication process regarding objectives and process (Dolinska and d'Aquino, 2016). After the initial phase, the approach opens when actions or initiatives were initiated, or when the need arises for consultation with a multiplicity of stakeholders with the aim of increasing ownership of the results. When the operating framework is a balanced mix of structure and flexibility, a dynamic movement between OE/D perspectives is set in motion, as see in the spiralling up and out dynamics described in the Italian case.

Finally, OE perspectives often tends to meet resistance because it is a vision that goes beyond the usual frameworks and therefore works in a disruptive fashion (e.g. disrupting power and knowledge relationships, work habits, etc.). The transformative potential is stronger than the disruptions, however; confronting the disruption can be beneficial given that it is associated with a collective reflexivity as well as trust and reassurance of partners through concrete shared and achieved actions. Conceptually, our study highlights the importance of focussing on trajectories, their dynamics in governance mechanisms, and the corresponding turning points as a motor of change in agri-food transition projects. We also underline the relevance of the "project" as category of analysis to investigate forms in which actors translate sustainability issues into objectives and processes. It takes goals and governance forms into account and from a dynamic perspective reveals the different visions and viewpoints of the actors while also showing the processes undertaken to reach (or not) a common horizon of commitments. Therefore, future research on agroecological transitions should pursue investigations of food transition projects in order to address the structures as well as the processes of governance. Our case study comparison showed how visions, expectations, perspectives and goals are translated at the territorial level into forms of governance that reflect existing political and power stakes. This chapter highlights that the formalization of transition perspectives into a "project" is a major challenge, particularly in the definition of objectives and collective processes. It is therefore crucial that these elements could be taken into account in the design of projects to adjust the speed and direction of the transition of food systems, as agroecological transitions often involve new governance mechanisms. Moreover, our case studies highlight that the creation of a mutual learning process enhancing collective reflexivity could help to collectively reorient and readjust the objectives, structure and processes of a transition process as confirmed by other researchers (Binder et al., 2015). Furthermore, reflexivity creates more realist expectations and clarifies the functioning and organization as actors simultaneously reflect about their role and the roles of other actors.

Finally, the territorial perspective seems particularly relevant for practitioners developing reflexive governance, as it is easier to consider stakeholder systems at the territorial scale. By examining the systems of actors via the same (territorial) perspective as the forms of governance, the inclusion and exclusion processes in transition projects could be explicit. The territorial perspective is therefore useful to implement in the practical actions of the diverse food systems' stakeholders, and in the analysis of the ecological and social processes favouring agroecological transitions (Lamine et al., 2019). In summary, our analysis via the lens of

OE/D processes and perspectives is useful for examining the ecological transition as revealed in the studied agri-food projects.

Acknowledgements

Our special thanks to Claire, Danièle and Terry for giving us the opportunity to work together on this chapter and get to know each other along the journey of drafts exchange and revision. We would like to express our gratitude to our case-studies key informants for allowing us to understand the dynamics of their territories and the interesting interplay between actors. For the Italian case study, thanks to Gianfranco Ciola, former director of the Coastal Dunes Regional Nature Park for his valuable support in data gathering and interpretation. The Provence Verte field-work was carried out with the support of the ADEME Agency, the Ecological Transition Ministry program Cit'In, and the cooperation of Provence Verte Agriculture High School (LEAP), which are hereby thanked. Also thanks to the Institute for Agriculture, Fisheries and Food Research (ILVO) in Flanders who supported the Flemish case study. The Mirecourt project was made possible by the financial support of Fondation of France. Moreover, we would like to thank the two anonymous reviewers for giving useful feedback that improved the quality of the book chapter. Thanks also to Miriam Levenson for English-language editing.

References

Binder, C. R., Feola, G., Steinberger, J. K., 2010. Considering the Normative, Systemic and Procedural Dimensions in Indicator-Based Sustainability Assessments in Agriculture, *Environmental Impact Assessment Review*, 30, 2, 71–81.

Binder, C. R., Absenger-Helmli, I., Schilling, T., 2015. The Reality of Transdisciplinarity: A Framework-Based Self-Reflection from Science and Practice Leaders, *Sustainability Science*, 10, 4, 545–562.

Borgatti, S. P., Molina, J. L., 2003. Ethical and Strategic Issues in Organizational Network Analysis, *Journal of Applied Behavioural Science*, 39, 3, 337–349.

Born, B., Purcell, M., 2006. Avoiding the Local Trap: Scale and Food Systems in Planning Research, *Journal of Planning Education and Research*, 26, 2, 195–207.

Bui, S., 2015. Pour une approche territoriale des transitions écologiques – Analyse de la transition vers l'agroécologie dans la Biovallée (1970–2015).

Claro, D. P., Hagelaar, G., Omta, O., 2003. The Determinants of Relational Governance and Performance: How to Manage Business Relationships? *Industrial Marketing Management*, 32, 8, 703–716.

Coquil, X., Anglade, J., Barataud, F., Brunet, L., Durpoix, A., Godfroy, M., 2019. TEASER-lab: concevoir un territoire pour une alimentation saine, localisée et créatrice d'emplois à partir de la polyculture – polyélevage autonome et économe. La diversification des productions sur le dispositif expérimental ASTER-Mirecourt, *Innovations Agronomiques*, 72, 61–75.

Dolinska, A., d'Aquino, P., 2016. Farmers as Agents in Innovation Systems. Empowering Farmers for Innovation through Communities of Practice, *Agricultural systems*, 143, 112–130.

Folke, C., Hahn, T., Olsson, P., Norberg, J., 2005. Adaptive Governance of Social-Ecological Systems, *Annual Review of Environment and Resources*, 30, 1, 441–473.

Garçon, L., Delfosse, C., Lamine, C., 2017. Disqualifier les labels pour requalifier produits, acteurs et lieux, *Systèmes alimentaires/Food Systems*, 2, 57–80.

Haberl, H., Gaube, V., Díaz-Delgado, R., Krauze, K., Neuner, A., Peterseil, J., Plutzar, C., Singh, S. J., Vadineanu, A., 2009. Towards an Integrated Model of Socioeconomic Biodiversity Drivers, Pressures and Impacts. A Feasibility Study Based on Three European Long-Term Socio-Ecological Research Platforms, *Ecological Economics*, 68, 6, 1797–1812.

Hubeau, M., 2019. The Potential of Agri-Food Networks in Food System Transformations. Doctoral dissertation. Ghent University.

Hubeau, M., Marchand, F., Coteur, I., Mondelaers, K., Debruyne, L., Van Huylenbroeck, G., 2017. A New Agri Food Systems Sustainability Approach to Identify Shared Transformation Pathways towards Sustainability, *Ecological Economics*, 131, 52–63.

Hubeau, M., Marchand, F., Coteur, I., Debruyne, L., Van Huylenbroeck, G., 2018. A Reflexive Assessment of a Regional Initiative in the Agri-Food System to Test Whether and How It Meets the Premises of Transdisciplinary Research, *Sustainability Science*, 13, 4, 1137–1154.

Klerkx, L., Leeuwis, C., 2009. Establishment and Embedding of Innovation Brokers at Different Innovation System Levels: Insights from the Dutch Agricultural Sector, *Technological Forecasting and Social Change*, 76, 6, 849–860.

Kloppenburg, J., Hendrickson, J., Stevenson, G. W., 1996. Coming in to the Foodshed, *Agriculture and Human Values*, 13, 3, 33–42.

Lamine, C., 2011. Transition Pathways towards a Robust Ecologization of Agriculture and the Need for System Redesign. Cases from Organic Farming and IPM, *Journal of Rural Studies*, 27, 2, 209–219.

Lamine, C., 2012. « Changer de système » : une analyse des transitions vers l'agriculture biologique à l'échelle des systèmes agri-alimentaires territoriaux', *Terrains & Travaux*, 20, 139–156.

Lamine, C., Garçon, L., Brunori, G., 2019. Territorial Agrifood Systems: A Franco-Italian Contribution to the Debates Over Alternative Food Networks in Rural Areas, *Journal of Rural Studies*, 68, 159–170.

Lamine, C., Magda, D., Amiot, M. J., 2019. Crossing Sociological, Ecological, and Nutritional Perspectives on Agrifood Systems Transitions: Towards a Transdisciplinary Territorial Approach, *Sustainability*, 11, 5, 1284.

Loorbach, D., Frantzeskaki, N., Avelino, F., 2017. Sustainability Transitions Research: Transforming Science and Practice for Societal Change, *Annual Review of Environment and Resources*, 42, 1, 599–626.

Marsden, T., 2000. Food Matters and the Matter of Food: Towards a New Good Governance? *Sociologia Ruralis*, 40, 1, 20.

Marsden, T., 2013. From Post-Productionism to Reflexive Governance: Contested Transitions in Securing More Sustainable Food Futures, *Journal of Rural Studies*, 29, 123–134.

Renting, H., Marsden, T. K., Banks, J., 2003. Understanding Alternative Food Networks: Exploring the Tole of Short Food Supply Chains in Rural Development, *Environment and Planning A*, 35, 3, 393–411.

Rossi, A., 2017. Beyond Food Provisioning: The Transformative Potential of Grassroots Innovation around Food, *Agriculture*, 7, 6.

Smith, A., Stirling, A., 2008. Social-Ecological Resilience and Socio-Technical Transitions: Critical Issues for Sustainability Governance.

Sonnino, R., Marsden, T., Moragues-Faus, A., 2016. Relationalities and Convergences in Food Security Narratives: Towards a Place-Based

Approach, *Transactions of the Institute of British Geographers*, 41, 4, 477–489.

Stotten, R., Bui, S., Pugliese, P., Lamine, C., 2017. Organic Values-Based Supply Chains as a Tool for Territorial Development: A Comparative Analysis of Three European Organic Regions, *The International Journal of Sociology of Agriculture and Food*, 24, 1, 135–154.

Turnheim, B., Berkhout, B., Geels, F., Hof, A., McMeekin, A., Nykvist, B., van Vuuren, D., 2015. Evaluating Sustainability Transitions Pathways: Bridging Analytical Approaches to Address Governance Challenges, *Global Environmental Change*, 35, 239–253.

Velly, R., 2019. Allowing for the Projective Dimension of Agency in Analysing Alternative Food Networks, *Sociologia Ruralis*, 59, 1, 2–22.

Voß, J. P., Kemp, R., 2006. Sustainability and Reflexive Governance: Introduction, *Reflexive Governance for Sustainable Development*, 25.

Yin, R. K., 2003. Case Study Research, in Maruster, L., Gijsenberg, M. G. (Eds.), *Design and Methods Applied Social Research Methods Series*, Thousand Oaks, CA, Sage Publications, 359–400.

How policy instruments may favour an articulation between open ended and deterministic perspectives to support agroecological transitions? Insights from a franco-brazilian comparison

CLAIRE LAMINE, CLAUDIA SCHMITT, JULIANO PALM,
FLORIANE DERBEZ, PAULO PETERSEN

1. Introduction

Even if the social structure of French and Brazilian agriculture is very different, in both countries, farming is at the heart of controversies and social struggles between its "dominant" forms and "alternative" or "differentiated" forms (such as "agribusiness" and "family farming" in Brazil), categories that are themselves quite heterogeneous. Connected to these controversies and social struggles surrounding economic, environmental and social issues, in both countries public policies have recently been forged based on the notion of agroecology. In Brazil, the strong involvement of civil society organizations allowed agroecology to gain space in the public policies, particularly in the ones related to family agriculture, from the mid-2000s on (Petersen et al., 2013). In France, the process was more government led and agroecology made its official debut in public policy in 2012 – although this had been anticipated by lasting debates in the agricultural world and civil society and reorientations in public agronomic research – when the then recently elected socialist government decided to put into place the frame of reference of agroecology, seen as an all-encompassing model (Lamine, 2017).

In both countries it has led to the elaboration of devoted policies and related instruments[1]. In Brazil, principles and objectives related to agroecology were incorporated to family farming policy instruments, coordinated by the MDA (Ministry of Agrarian Development[2]) and to wider policies articulating agriculture, food, health and environment, implemented in partnership with different Ministries (Schmitt 2016), such as the Food Acquisition Program and the National School Feeding Program[3]. In 2012 a National Policy of Agroecology and Organic Production (PNAPO) was launched, and translated into a national action plan also creating a favourable environment for building state-level agroecology policies (Guéneau et al., 2019). The construction of the PNAPO was a major step in structuring governance mechanisms capable of streamlining the incorporation of agroecology (principles and guidelines) into a diverse set of policy instruments, with the involvement of social movements and representatives of civil society organizations. One of the most innovative of these policy instruments, on which we will focus here, is the Ecoforte program, aimed at supporting projects to be developed at a territorial scale by networks of organizations working with agroecology, extractivism and organic production, in order to support the dissemination of practices related to the sustainable management of biodiversity and to organic and agroecological farming systems.

In France, the national agroecological policy was launched in 2012 and named *Produire Autrement* ("produce differently"), and the government proposed a legislation promoting "the economic and environmental dual-performance" of agriculture (the law was passed in 2014). Soon after this national policy was launched, a first policy instrument was set up, through a call entitled *Mobilisation Collective pour l'Agro-Ecologie* (Collective Mobilization for Agroecology, hereafter MCAE), in May 2013, aimed at supporting farmers or multi-actors groups that would develop agroecological approaches at the territorial scale. This policy

---

[1] Policy instruments can be described as technical and political devices that organize the relationships between public authorities and their target audiences based on the representations and meanings they embody (Lascoumes and Les Galès, 2004). In this chapter we study two policy instruments or "programs" that take the form of public calls for tender that fund projects for a limited period of time (2–3 years).

[2] Created in 1999 and devoted to family agriculture, and suppressed in 2016.

[3] At least 30% of its resources should be invested in the purchase of family farming products. Local products, products grown by specific groups (land reform settlers, indigenous people and traditional communities) and organic or agroecological products should be given priority.

instrument was conceived as experimental within a a larger agroecological plan articulating diverse on-going and new programs, tools and reforms to support the agroecological transition of French agriculture (such as pesticide reduction, seed evaluation, investment subsidies, diagnostic tool for farmers, agricultural education etc.)[4].

How do these policy instruments cope with the characteristics of agroecological transitions (AET)? It is widely acknowledged that AET are diverse and "situated," depending on contexts; systemic (relying on and bringing together ecological processes but also social ones); and that they require flexibility and adaptive perspectives (Elzen et al., 2017; Ollivier et al., 2018), as is explored in other chapters of this book. Hence, policy instruments aimed at supporting agroecological transitions (AET) should take into account these characteristics, in a context where policy instruments in the agricultural field usually focus on techniques (to the expense of more systemic visions), on the individual scale (as opposed to collective and/or territorial and/or multi-actors scales), and adopt deterministic perspectives on trajectories of change, whereby the target and goals should be defined (in contrast to "*in itinere*" definition).

MCAE and Ecoforte, in contrast to more conventional policy instruments such as agri-environment schemes, input reduction plans or subsidies for organic conversion, are innovative for three main reasons: their target is not individual farmers but groups with a multi-actor approach trying to involve not only farmers also other food systems actors; their approach is territorial; and they allow the actors to build their own trajectory of change. **This last characteristic is linked to both their *combining* a deterministic and open-ended perspective to AET in the way they are conceived and through the frame they provide to the groups' projects (normative effect), and their *allowing a articulation* of these two perspectives through the way the groups use these instruments and adapt their potentialities to their own situation (performative effect).** These two points will constitute the two steps of our demonstration. The first one is explicitly aimed at and formalized in the policy instruments themselves through their conception, the calls and the evaluation procedures, while the second one is less explicit and obvious. By favouring this combination and articulation between open ended perspectives to AET and more

---

[4] https://agriculture.gouv.fr/agroecology-project-france. Accessed on the 20th of July 2020.

deterministic ones, such policy instruments better adapt to the need of contextualization and adaptability that characterizes AET[5].

2. Analytical framework

The construction and implementation of agroecological policies and instruments in France and in Brazil have given rise to recent works that have attempted to characterize these policies and instruments in contrast to previous ones (Flexor and Grisa, 2016; Schmitt et al., 2017; Lamine et al., 2019). This has led some authors to highlight the persistent productionist stance within current agroecological policies (Arrignon and Bosc, 2017; Rivera-Ferre, 2018), as was the case with previous policies and policy instruments such as the agri-environmental schemes (Evans, 2009). This echoes the lasting criticism that has long been developed about agricultural policies, which some consider as framed by certain (mainstream) actors and visions (Muller, 2000), despite the process of ecologisation of agricultural public policies which started in Europe in the mid 1980s (Deverre and De Sainte Marie, 2009) and was institutionalized under the 1992 CAP reform (Fouilleux, 2000). This ecologisation of agricultural public policies has thus been criticized for being more "green-washing" than transformative, because of their normative and "deterministic" stance (Mormont, 1996). Another weakness is that they do not allow for coordinated changes at the scale not only of farms, farmers' practices and agroecosystems but also of agrifood systems because they lack a systemic and encompassing vision, do not include upstream economic actors such as inputs providers and downstream ones like processors, retailers, consumers, nor address the necessary changes within the agricultural knowledge and innovation system (Lamine, 2011). In this context, many authors today argue in favour of more transformative policies (Levidow, 2015; Marsden et al., 2019).

Our approach is anchored in a pragmatist perspective, whereby we analyse in an articulated manner the visions and discourses held by the actors involved in these projects as well as the debates and controversies at play, and the concrete projects and actions they carried out within these specific programs. Our focus on what these policy instruments perform (Lamine et al., 2020) allows us to escape the classical alternative

---

[5] The research took place in two national projects mentioned in this chapter, and the comparison between the two instruments, within the Capes Cofecub project SH 944/19 on AET.

between a "critical sociological" perspective aimed at unveiling the persistent productionist stance, or the naive valuation of localized and non reproductible initiatives. Our analysis is also anchored in an action research perspective, as these two policy instruments have been studied based on two innovative collaborative "observatories" involving local actors, institutional actors and researchers, with the key principle to elaborate and discuss our findings with these actors.

In the case of MCAE, the ObsTAE project (Sociological Observatory of AETs, 2015–2018) gathered 12 researchers and PhDs in social sciences in both an analytical work on the trajectories of the supported projects and a participatory approach structured through cross-projects multi-actors seminars. In the case of Ecoforte, the project Agroecology Networks for the Development of Territories (ANA/FBB/BNDES – 2017–2018), coordinated by the Executive Secretariat of the National Alliance of Agroecology (ANA), with strong involvement of the Ecoforte-supported networks, involved a participatory process of systematization of processes and results[6] generated within the Ecoforte Program.

Our analytical grid is composed of two main parts, aimed at informing first the normative effects of these instruments, based on the analysis of the way they were conceived, the objectives and criteria expressed in the calls and their implementation process (evaluation and selection) and thus the frame they provide to the farmers or multi-actors groups that apply to these calls; and second their performative effects, based on the analysis of the way the groups appropriate and use these instruments and adapt their potentialities to their own situation.

Both aspects required a combination of methods (documentary analysis, interviews, participant observation) – see Tab. 1. For the French case, a three years qualitative and collaborative study and follow-up of 16 projects was carried out, based on interviews (10–20/project), participant observations, and five cross-projects multi-actors seminars at the national

---

[6] In Latin America, the systematization of experiences is a practice of participatory research that emerged in the pedagogical movement of "popular education" and aims to produce knowledge, with the involvement of social actors, through integrated cycles of action and reflection. It contemplates a plurality of epistemic perspectives and methodologies but can be characterized in broad lines as an effort to orderly reconstruct collective experiences, from a critical perspective, seeking to understand its conditionings, and favour collective learning (Holliday, 2006; Cordero and Carrillo, 2017).

scale. For the Brazilian case, the participatory research encompassed three levels of analysis involving 25 networks out of the 28 supported by the call: (1) these 25 proposals as well as the accountability reports were analyzed based on a list of questions developed by the systematization team; (2) an assessment of the actions developed by 15 networks was based on short-term field work in the territories; (3) a deeper reflection exercise was carried out about the actions developed by three selected networks, based on in-depth studies on specific themes chosen by the networks' actors (Schmitt et al., 2020).

In both cases, the studied projects were chosen in order to contemplate a variety of situations from the point of view of their location, social composition, time of existence, type of ecological agriculture at stake.

Tab. 1: Data analyzed for each instrument.

	MCAE	Ecoforte
Analysis of the call documents	Public call 2013	Public call 2014
Analysis of the construction process	Participation to meetings, minutes	Interviews and documents
Analysis of the selection process	Final results (projects abstracts)	Final results and interviews with public agents participating in this process
Analysis of the projects	Projects abstracts (469 submitted, 103 selected) In depth analysis of 16 projects	In depth analysis of the 28 selected projects.
Analysis of the on-going projects	16 projects analyzed through interviews and participant observation	25 projects included in the participatory process of systematization.

3. Two innovative tools which focus on collective, multi-actors and territorial scales and dynamics

3.1. The Collective Mobilization for AgroEcology Call (MCAE)

This call was in 2013, the first specific policy instrument of the French agroecological policy and aimed at financing over three years, farmers or multi-actors groups to support their AET at the collective and territorial scales and was conceived as an experimental platform for the future Environmental and Economic Interest Groups (GIEE; a new legal entity type that was then in construction, and represents about 500 groups in 2020). Most of the 103 groups that were funded through MCAE between 2013 and 2017 indeed chose to transform into GIEEs (Tab. 2).

The program aimed at supporting « *ascendant territorial collective approaches* », seen as « *complementary to more traditional descendant approaches* » and « *forms of agriculture with a high economic and environmental performance* »[7], in a *"non-normative"* way. The bottom-up nature of projects was indeed one of the key criteria of evaluation. The call was also innovative in that is was open to farmers groups as such, while usually government funds would only be attributed to development and extension "intermediary" organizations. Another innovation was the possibility for farmers belonging to these groups, to be refunded for the time they would spend in training their colleagues or managing the projects.

The call was prepared by a conception working group composed of officers of the Ministry of Agriculture, public agencies or foundations, and researchers. As one of the main challenges was to open the range of projects and organizations that could be supported, beyond the institutional and mainstream actors of agricultural development, and because some participants in the conception group had warned that institutionalized agricultural organizations were much more used to setting up such projects than smaller and more alternative ones, the evaluation process was conceived in order to avoid a stranglehold by the incumbent actors and to include a diversity of organizations. Multi-actors evaluation committees were set up at the regional and national scales, that involved experimented agricultural development agents, civil society organizations, local authorities representatives, government officers and researchers. This opened the way to quite a diversity of profiles among the beneficiaries. Among the 103 beneficiary groups, besides the traditional actors of agricultural development (chambers of agriculture and cooperatives), we find less traditional entities, such as farmers groups that were involved in diverse forms of ecological agriculture, from conservation agriculture to organic farming, alternative agricultural and

---

[7] Circulaire DGPAAT/SDDRC/C2013-3048, 7 May 2013 (MCAF call)

rural development organizations, original partnerships between farmers and municipalities, as well as newcomers in the landscape of agricultural development such as specialized consulting groups.

Regarding the 16 collectives on which we focused, a common methodological framework and shared format of reporting was elaborated in order to enable comparisons of groups' dynamics. This allowed us to describe the emergence of these collectives as well as their trajectories, to analyse their objectives, the actions that had been carried out, their internal organization and external partnerships, as well as their visions and definitions of agricultural change and of agroecological or ecological practices. Among these 16 groups and projects, we focus here more specifically on two of them, both situated in southeast France. The first one is Amap, a regional network of community-supported agriculture groups that set up this project in order to foster a specific and encompassing vision of the agricultural activity including also relations with consumers, based on training sessions, the definition of appropriate indicators for their specific model, and an innovative tutorship system. The second was Addear 42 (Association for the Development of Agricultural and Rural Employment of the Loire *département*), which is linked to the *Confédération Paysanne* peasant union, and structured a project in order to support a group of farmers involved in maize seeds production and exchange and the collective testing of varieties adapted to local needs.

3.2. Ecoforte

The Ecoforte Program was launched in 2014, through a public call[8] published under the coordination of the Banco do Brasil Foundation (*Fundação Banco do Brasil* – FBB) and the Brazilian National Bank for Economic and Social Development (*Banco Nacional de Desenvolvimento Econômico e Social* – BNDES), the two institutions responsible for the direct coordination of the program and in charge of the funds allocated in this call[9]. The formulation of this program mobilized two management bodies created under the PNAPO: the Interministerial Chamber

---

[8] Call for Public Selection n° 2014/005. https://fbb.org.br/pt-br/viva-voluntario/conteudo/edital-de-selecao-publica-n-2014-005-redes-ecoforte. Accessed 22/11/2019.

[9] As a legal entity under private law, FBB was able to operate through more flexible formats in contracting projects with social organizations, in comparison to Federal

of Agroecology and Organic Production (CIAPO), composed of different state agencies, and the National Commission of Agroecology and Organic Production (CNAPO), an arena for dialogue and social participation bringing together government representatives and civil society organizations for the elaboration and monitoring of the National Policy[10]. Like MCAE, this program was formulated by a technical group, counting with the participation of representatives of 11 different institutions and of the National Articulation of Agroecology (ANA). The political and institutional bond of this group to both CIAPO and CNAPO undoubtedly expanded its political capacity in the implementation of Ecoforte.

According to Ecoforte rules, proposals could be submitted by networks working with agroecology, organic agriculture and sustainable extrativism, and should involve at least three producers' associations or cooperatives. Project activities should benefit family farmers, land reform settlers, indigenous people and traditional communities. This new policy instrument embodied into its regulatory framework a series of original concepts and procedures: projects should be carried out through a network of social organizations and incorporate a territorial approach (albeit with very flexible margins) and present a work plan oriented towards the implementation of an integrated set of *"reference units."* A reference unit was defined as *"a place of installation or demonstration of techniques, processes, methodologies or production systems where visits, exhibitions and training sessions are developed in order to promote the exchange of knowledge and the dissemination of experiences."*[11] Among the items to be financed through Ecoforte were tangible investments (including machinery, equipment, vehicles and facilities), as well as intangible ones (such as technical assistance or educational and training activities).

Government agencies. In the case of BNDES, non-repayable resources managed by specific funds (such as the Social Fund and the Amazon Fund) could be allocated to this program.

[10] These arrangements for interministerial coordination and civil society participation implemented by PNAPO were extinguished by Decree 9.784/2019, which aimed to dismantle a wide range of institutional arenas of social participation related to the formulation and monitoring of public policies in Brazil, in a larger context of weakening of the democratic principles established by the 1988 Constitution.

[11] Call for Public Selection nº 2014/005, *op. cit.*

The scoring system that guided project selection (carried out by a technical group from FBB and BNDES) valued the network's historical trajectory and its experience in implementing government programs and actions in similar areas (agroecology, organic production and sustainable extractivism), as well as the diversity of entities participating in the network and the involvement of women, youth, indigenous peoples and traditional communities. A total of 166 networks participated in the public announcement, of which 33 were approved and 28 effectively financed (Martins, 2018).

Among the 25 projects that were part of the systematization effort, two cases will be discussed here in more detail: the Ecovida Agroecology Network in the state of Rio Grande do Sul, involving the production, processing, certification and commercialization of sociobiodiversity products, and the network "Sabor Natural do Sertão" in the semiarid region of the state of Bahia, focused on access to water and the marketing of local products.

Tab. 2: Characterization of each instrument.

	MCAE	Ecoforte
Dates	2013–2017	2013–2017
Funding	6,5M€ total funding Max 100.000€/project	45,1MR$ (+/- 10M€) Max R$ 1.250.000,00/project (+/- 250.000€)
Number of submitted and funded projects	469 submitted 103 funded	166 submittetd 28 funded
Implementation	Ministry of Agriculture and its regional administration	FBB and BNDES

4. From the framing of AET by policy tools to their enactment by situated collectives

The focus here is to explore which kinds of AET these policy instruments aim at and perform. For this, we are going to analyse for each of them, (i) how agroecology and AET are framed and defined in the call; (ii) how they are defined by the groups; (iii) what these policy instruments perform through what actors make of them "in situ."

4.1. MCAE: An open definition of agroecology and agroecological transitions that gives way to a diversity of pathways… (and controversies)

In the MCAE call, agroecology was defined in a (claimed) open way, through key principles such as systemic approaches and the notion of "*double performance*" (economic and environmental one) which was key to the government's discourse at that time and was later extended to the notion of triple performance (including social one) following the intense controversies that developed between the government's vision of agroecology and that of social movements (Lamine, 2017). In the call text, some examples of agroecological practices were suggested to exemplify the types of innovations that could be implemented, such as soil conservation, feed autonomy, crop diversification, methanisation etc.

Agroecological transition was not mentioned as such, but the call aimed at supporting "*territorial collective bottom-up approaches in favor of agroecology*" (call, p2). Projects were expected to set up and disseminate agroecological practices and innovations, whether technical, organizational or social, contributing to improve the "*double performance.*" The groups were expected to define their own objectives and indicators in terms of actions and results achieved, which expresses the non-normative way put forward in the agroecological program, in contrast to more conventional instruments with pre-formated list of objectives and practices.

The *selection criteria* focused on the "double performance" aspects, on the reality of the farmers' involvement and the collective dynamic at stake, on the bottom-up nature of projects, on the partnership and the adaptation to the territorial context and needs.

Based on this framework, what did the groups that submitted a project aim at? The lexical analysis of the 469 submitted projects and our 16 qualitative studies showed that agroecology as such was not often mentioned in the text of these projects. When it was, this was often to "tick the good boxes" of the call. The projects rather mentioned the agricultural models previously adopted by the groups, such as organic agriculture, peasant agriculture, conservation agriculture etc. Some groups would claim a "natural" convergence of their already legitimate models (such as organic farming) with agroecology. This was the case of the Amap and Addear groups that both put forward the fact that they were already practicing agroecology without calling it this way and well before the government discovered and institutionalized it (which also echoes the controversies mentioned above, as social movements first of all criticized a re-appropriation of agroecology by the goverment): *"When we applied to the call, we thought, well is not it agroecology that we are already doing, we have a systemic approach, we already were in agroecology, a modest and peasant form of it."* (Addear, MCAE multi-actors seminar, May 2015). The Amap define themselves as « *precusors* » of agroecology, for two reasons. First, they claim to already have the triple economic, environmental and social performance – social dimensions are de facto central to their national charter and their actions, through the focus on producers-consumers links and sustainable livelihood. Second, they claim to already have a systemic vision, one that reconnects agriculture and food: *"Agroecology is what we claim when we say that agriculture has to be reconnected to food"* (Amap meeting, November 2015). By contrast, in other groups, agroecology appeared as a convenient narrative for "intermediate" models in search of legitimacy, such as soil conservation agriculture.

AET transitions were defined in very diverse ways in the projects, with goals and steps more or less defined and thus more deterministic or more open ended perspectives. In some groups, precise targets are defined for farmers trajectories in terms of agricultural practices (and their impacts), along with precise steps, and also for the group itself – as the collective activities are supposed to facilitate these trajectories at the farmers' scale. This is for example the case for some projects in conservation agriculture – partly because these groups have to "prove" the reality of their ecologisation processes. In other groups, the objective that are defined are much more processual than technical: it may be to foster exchanges about practices, to share seeds, to provide farmers with tools and tricks regarding non technical issues such as work organization or marketing issues. Some claim to (re)define the objectives along the way.

For the Amap, the project was not primarily focused on the ecologisation of farmers practices: most of them were already organic, or at least used no chemical inputs (even though AET might develop within organic agriculture, for example in order to reduce the dependence to external inputs). The objective was rather the dissemination and the viabilisation and stabilization of an original agroecological model including, beyond agricultural practices, work organization, relations to consumers, marketing issues. However, in order to be eligible for the call, the network had to frame its project under the agroecological "banner" and included the notion of *"optimization of agroecological practices"* in the title along with a core and peculiar claim which was the *"spreading of a sustainable agricultural 'handicraft' (artisanal) model."* The MCAE funding allowed the network to undertake a deep investigation about the trajectories, successes, failures and difficulties of its farmers, as well as a collection of innovative technical and non technical practices that could be adapted by other farmers in the Amap networks. This gave place to the production of a "guide of good pratices" – again a notion aimed at adapting to the funders' requirements – and of farmers' portraits that aimed at retracing the multi-dimensionality of their activities and the key factors of success.

Finally, **how did the groups appropriate and use this instrument and adapt its potentialities to their own situation**? Generally speaking, our ethnographic observation allowed us to assess a lot of **readjustments of trajectories** during the 3 years (i.e., within an adaptive vision). Most groups, in our seminars, highlighted the relevance of this instrument and the fact that it fitted their own ways to support farmers' AET, by prioritizing the actions based on the analysis of farmers' needs. In the Addear 42 and according to one of its facilitator, MCAE was considered as a very innovative and powerfull instrument in comparison with the tools they used to mobilize (mostly agricultural training funds) because they could experiment a "long term" support (in comparison to short and occasional training courses) and more participatory modes of governance which contributed to a (re)adjustement of their objectives throughout the duration of the project, especially regarding on-farm experimentation in order to adapt to the group's evolving needs.

4.2. Ecoforte: Expanding and contextualizing agroecological visions

The Ecoforte call was focused on the notions of "agroecological-based production," "organic production" and "extractivist production," all notions that have been defined and stabilized by a series of legislations through an ample process involving state and civil society actors[12]. This reference to well established legislations also contributed to the legitimacy and the legality of Ecoforte's actions. This broad view of agroecology aimed at taking into account the actual diversity of ecological models

---

[12] Law 10.831/2003 that institutionalizes organic agriculture in Brazil; Law 11.326/2006 that legally recognizes Family Farming and Rural Family Enterprise;

among the rural communities that may apply to the call. "Agroecological-based production" was defined as *"one that seeks to optimize the interaction between productive capacity, use and conservation of biodiversity and other natural resources, ecological balance, economic efficiency and social justice,"*[13] and not restricted to organic farming. Extractivist production was defined in relation to the sociobiodiversity products[14] and to production chains linked to family farmers. The reference to family agriculture relies on a rather large definition, and includes farmers, extractivists, small fishers etc., once again with the idea to include the diverse identities at stake.

Agroecological transition was defined in the call as *"a gradual process of changes in practices and management of traditional or conventional agroecosystems, through the transformation of the productive and social bases of the use of land and natural resources, which lead agricultural systems to implement ecological principles and technologies."*[15] This rather wide definition of AET as a social and ecological process of transformation, that also refers to the need to adapt to the diversity of contexts, breaks with a technicist vision of agricultural changes that focuses on the sequential incorporation of a defined set of technologies. It also distances itself from a focus on productive units, as the intervention scope of the program encompasses a large diversity of interactions and processes involved in the transformation of not only productive systems but also larger agrifood systems. The Ecoforte's definition of AET includes the social, cultural and identity components as being a full part of transition processes.

What did the groups that submitted a project aim at? The projects were to be designed as a set of *reference units*, to be implemented by

---

PNAPO's regulation Decree (Decreto n° 7.794/2012) and the Interministerial Ordinance creating the National Plan for the Promotion of Sociobiodiversity Production Chains.

[13] Call n° 2014/005, *op. cit.*

[14] "Sociobiodiversity products" are defined as goods and services derived from biodiversity and inscribed in productive chains of interest to traditional people and communities and market circuits that value peoples' knowledges and practices, promote their rights, incomes and quality of life. The notion of sociobiodiversity aims at highlighting the social and cultural aspects associated with the reproduction of biological diversity (Viana, 2015).

[15] Call n° 2014/005, *op.cit.*

a particular network. Their activities were meant to contribute to the strengthening of the network as a collective device, fostering AETs. The concept of reference unit was translated in different ways in the studied projects, encompassing demonstrative vegetable gardens and orchards, animal breeding, agroforestry systems, locally adapted seed banks, food processing units, water catchment systems, multipurpose training centres etc. In the case of the network *Sabor Natural do Sertão*, located in a semiarid region in Bahia, access to water is a crucial factor. Water collection and storage technologies allowed to intensify home garden production as well as women's participation, enabling the establishment of integrated vegetable and livestock systems. The project also supported the structuring of *Armazém da Central*, a store established in the city of Juazeiro, which now displays over 300 products from 30 cooperatives. Regarding the Ecovida Agroecology Network in Rio Grande do Sul, the project sought to structure a series of activities aimed at the production, processing and consumption of native fruits, supporting different activities, including training, technical support, marketing, certification and others. Ecovida Nucleos operating in different biomes became more integrated through the project.

How did the groups appropriate and use the Ecoforte instrument and adapt its potentialities to their own situation? The aim of the program was that each network could establish its own strategy, involving reference units as defined above. The analysis showed that various strategies were used by these networks to strengthen their action at the territorial level: increasing the number of farmers and extractivists involved in agroecology and organic production, developing new products and activities, strengthening participation in different markets, intensifying activities with specific groups (including young, women and traditional peoples and communities), mobilizing public policies beyond Ecoforte. In Ecovida Network, the dynamics established within the "Meeting of Flavors" (*Encontro de Sabores*), a processing and storage unit, generated a series of new customer relations flows through fairs, gastronomic events, stores, demanding adjustments in order to allow a better balance between supply and demand and a greater diversity of products. These adjustments were made possible by the flexibility of the Ecoforte Program. Our analysis of the projects also shows an increased capacity of these networks to mobilize a broader set of public policies, taking advantage of the new capabilities and opportunities driven by Ecoforte.

5. Discussion

Our comparison shows many commonalities between the two instruments in their wish to favour an open-ended perspective to AET. This allowed funded projects to design their own transition pathways and their own combination of open-ended and deterministic perspectives.

5.1. Scope, actors and scale as key objects of the open-ended perspective claimed by both programmes

The definition of the scope, the identification of the actors to be involved, and the choice of the relevant scale appear as three key elements that translated the claim for an open-ended perspective to AET.

Both MCAE and Ecoforte put forward the multi-dimensionality of AET, in that not only agricultural practices but also rural development, reconnexion between agriculture and food, and social inclusion were to be addressed by the projects. However, this reconnexion was more effective in the Brazilian case thanks to the longer experience of institutionalisation of intersectoriality, that is, of the articulation between sectoral policies. In France by contrast, some projects remained focused on agricultural practices alone, with a weak articulation with food practices, territorial development and social issues.

The wish to widen the range of actors to be involved (beyond farmers and technical advisors) was present in both cases, both at the stage of the construction of the instrument (the working groups and commissions that worked on the design of these two instruments involved a large range of actors) and in the rules set up in the calls, that also required the involvement of a large range of actors in the projects to be financed.

Finally, the definition of the relevant scale was also a key issue in both calls, and the territorial scale was in both cases favoured to the farm scale or the food chain one. Both also admitted a flexibility and an "openness" in the definition of this scale[16]. For exemple, the Ecoforte worked with a flexible definition of territory, generally defined as the operating space of the network and its organizations.

---

[16] In this way, the projects were able to express different scale strategies, related to the contextualized processes of AET.

5.2. Modes of articulation of deterministic and open-ended perspectives

This claim for a more open ended perspective is expressed in the fact that both programmes allow farmers' groups to define their own trajectory, to redefine steps and actions along the way, and to design pathways and developments that are adapted to their specific contexts. This is in contrast to having to adapt a predefine scheme for example, as is often the case in more conventional policy instruments such as agri-environmental schemes.

However, our analysis of the rules of the two programs and of the processes of appropriation by diverse groups, shows that beyond this shared claim of an open ended perspective, these programmes actually rather require an *articulation or a combination* between such an open ended perspective and more deterministic ones. Many rules in these calls – as in public calls in general – lead to reduce the openness, as in the French case were groups had to report at the end of the project in a pre-defined way that did not always fit with the reality of the dynamics they had set up during the three years. Reversely, some groups had pre-defined in their projects goals and pathways and thus adopted a more deterministic perspective than others, but then de facto adapted their actions and steps along the way and finally acknowledged the flexibility of the instrument. It is precisely this feature of the two programmes – to allow for a combination of these two perspectives, with diverse balance between the two, that supports the adaptation to the groups' contexts.

5.3. The role of researchers in the redefinition of "systematization"

Both instruments aimed at a larger diffusion and adoption of the innovative approaches that were designed and implemented by the funded projects. Therefore they had strong expectations in terms of both *"systematisation"* (in France, *capitalization,* a key word of the agroecological policy) and *dissemination* of these innovative approaches to other groups and regions. This has raised a key paradox between these public policies' expectations and the fact that the knowledge that is produced is intrinsically situated, that is, anchored in the groups' specific geographical, agronomical and social contexts. In both cases, the action-research approach led to argue for the production of *contextualized* learnings that can be *adapted* rather than *transferred* to other contexts.

The question of how to report on the reality and the diversity of AET, given this context-dependency, is indeed a key challenge for sociological analysis. In both cases, the social scientists carried out a collective inquiry that involved the concerned actors. Its outcomes were shared with the diverse actors involved in these programmes and projects, within regular workshops and through larger events bringing together a wider range of actors (agricultural and development organizations, policy makers, researchers). The extent to which this transdisciplinary process, which combined sociological observation with multi-actors arenas favouring "collective intelligence," could also generate learning processes for the policy makers is still to be assessed[17].

6. Conclusion

Besides their collective, multi-actors and territorial focus, the innovative dimension of these two experimental instruments lies in the combination of deterministic and open-ended perspectives that they allow, as was shown based on (i) the analysis of the frame they establish for the collective projects they fund (through the calls, rules, criteria etc.) – their normative effects; and (ii) the analysis of the ways they are appropriated by the groups – their performative effects. In both cases, the multi-actors seminars and collective work around these projects converged to praise this principle of combination in agroecological policies.

What are their actual concrete effects in terms of AETs at the scale of the farms, groups and territories involved was not the focus of this article, but in both cases the groups were able to implement a series of strategies of scaling-up and scaling-out, incorporating new farmers, strengthening participation in different markets, expanding the diversity and the ecological complexity of their production systems. In Brazil, the application of the Lume method of economic and ecological analysis of agroecosystems[18] allowed to show that the metabolic profile (use of energy

---

[17] In the Brazilian case, these collective learning processes were taken into account in the formulation of the second edition of the call (in 2017).

[18] The Lume Method was developed by the ONG AS-PTA – Family Farming and Agroecology, affiliated to the ANA. For a detailed assessment of the Lume method see: Petersen et al., 2017.

Policy instruments for the agroecological transition 147

and material) of local food systems at farm scale was reduced thanks to the organization of territorial markets, collective processing units, and mechanisms to manage shared resssources (Schmitt et al., 2020).

The incorporation of more open-ended perspective on AET, departing from the more linear definitions of "conversion to organic farming," as well as that of network, was the result of a long trajectory of social criticism and militancy. This enabled that social and organizational knowledge developed within civil society networks could be translated, at least to some extent, to public policies (Swako and Lavalle, 2019), as was the case in Brazil between 2003 and 2016.

However, these programs remain fully anchored in the "projectification" or "project city" mode (Boltanski and Chiapello, 1999; Sjöblom and Godenhjelm, 2009) which intrinsically tends to frame transitions in rather deterministic ways, with pre-defined objectives and steps or milestones and supporting transition processes over short periods of time. Despite the calls were open to farmers' groups and networks, specific skills that are necessary to design and write such projects are of course not accessible to all farmers' groups, and this frames the range of social groups that can benefit from these instruments. In the Brazilian case the financial accountability system was perceived as being quite demanding. In the French case, some « business-as-usual » processes of appropriation were observed, due to the lasting power relationships between conventional and alternative agricultural actors. This "reframing"/"reclosing" process has been limited by the role played by the MCAE national evaluation committee whose composition had been carefully thought in order to escape these power inequalities, but with the later GIEE evaluation processes, we observed diverse degrees of "reframing" in the regional commissions that were in charge of evaluating the projects (Lamine et al., 2020). Besides these unequalities in terms of access to public funds, another limit lies in the discontinuity linked to political changes, especially in the Brazilian case where many programs that supported the various actions undertaken by the groups were stopped in the recent period (even though Ecoforte itself was relaunched). The assessment and understanding of longer term effects would require lasting "observatories" of which the two projects carried out around these two instruments were only prototypes.

Acknowledgements

We would like to thank the two anonymous reviewers for their useful suggestions as well as the Capes Cofecub project SH 944/19 for supporting this comparative research, the French Ministry of Agriculture and the Brazilian FBB and BNDES for supporting the national studies.

References

Arrignon, M., Bosc, C., 2017. *Le plan français de transition agroécologique et ses modes de justification politique*. https://www.cairn.info/les-politiques-de-biodiversite--9782724621709-page-205.htm.

Boltanski, L., Chiapello, E., 1999. *Le nouvel esprit du capitalisme*, Paris, Gallimard.

Cordero, D. B., Carrillo, A. T., 2017. *La sistematización como investigación interpretativa crítica*, Bogotá, Editorial El Burro/Corporación Síntesis.

Deverre, C., De Sainte Marie, C., 2009. L'écologisation de la politique agricole européenne. Verdissement ou refondation des systèmes agroalimentaires? *Revue d'Etudes en Agriculture et Environnement*, 4, 89, 22.

Elzen, B., Augustyn, A. M., Barbier, M., van Mierlo, B., 2017. AgroEcological Transitions: Changes and Breakthroughs in the Making. DOI: http://dx.doi.org/10.18174/407609.

Evans, N., 2009. Adjustment Strategies Revisited: Agricultural Change in the Welsh Marches. *Journal of Rural Studies*, 25, 2, 217–30.

Flexor, G., Grisa, C., 2016. Contention, Ideas, and Rules: The Institutionalization of Family Farm Policy in Brazil, *Canadian Journal of Latin American and Caribbean Studies/Revue canadienne des études latino-américaines et caraïbes*, 41, 1, 23–37. https://doi.org/10.1080/08263663.2015.1130292.

Fouilleux, E., 2000. Entre production et institutionnalisation des idées. La réforme de la Politique agricole commune, *Revue française de science politique*, 50, 2, 277–306. https://doi.org/10.3406/rfsp.2000.395468.

Guéneau, S., Sabourin, E., Colonna, J., de Freitas Ewald Strauch, G., Piraux, M., Lamine, C., Lucio de Ávila, M., et al., 2019. A construção das politicas estaduais de agroecologia e produção organica no Brasil, *Revista Brasileira de Agroecologia*, 14, 2, 7–21. https://doi.org/10.33240/rba.v14i2.22957.

Holliday, O. J. 2006. *Para Sistematizar Experiências*, Brasília, Ministério do Meio Ambiente.

Lamine, C., 2011. Transition Pathways towards a Robust Ecologization of Agriculture and the Need for System Redesign. Cases from Organic Farming and IPM, *Journal of Rural Studies*, 27, 2, 209–19. https://doi.org/10.1016/j.jrurstud.2011.02.001.

Lamine, C., 2017. *La fabrique sociale de l'écologisation de l'agriculture*, Marseille, La Discussion.

Lamine, C., Niederle, P., Ollivier, G., 2019. Alliances et controverses dans la mise en politique de l'agroécologie au Brésil et en France, *Natures Sciences Societe*, 27, 1, 6–19.

Lamine, C., Barbier, M., Derbez, F., 2020. L'indétermination performative d'instruments d'action publique pour la transition agroécologique, in Arrignon, M., Bosc, C. (Eds.), *Les transitions agroécologiques en France – Enjeux, conditions et modalités du changement*, Presses Universitaires Blaise-Pascal, PUR, 37–52.

Lascoumes, P., Le Gales, P., 2007. Introduction: Understanding Public Policy through Its Instruments—From the Nature of Instruments to the Sociology of Public Policy Instrumentation, *Governance*, 20, 1, 1–21. https://doi.org/10.1111/j.1468-0491.2007.00342.x.

Levidow, L., 2015. European Transitions towards a Corporate-Environmental Food Regime: Agroecological Incorporation or Contestation? *Journal of Rural Studies*, 40, 76–89. https://doi.org/10.1016/j.jrurstud.2015.06.001.

Marsden, T., Moragues Faus, A., Sonnino, R., 2019. Reproducing Vulnerabilities in Agri-Food Systems: Tracing the Links between Governance, Financialization, and Vulnerability in Europe Post 2007–2008, *Journal of Agrarian Change*, 19, 1, 82–100. https://doi.org/10.1111/joac.12267.

Martins, J. M. R., Sambuichi, R. H. R., 2019. Programa Ecoforte e o fortalecimento das redes de agroecologia: demandas e possibilidades, Texto para Discussão, 2455, Rio de Janeiro, IPEA.

Mormont, M., 1996. Agriculture et environnement : pour une sociologie des dispositifs, *Économie rurale*, 236, 1, 28–36. https://doi.org/10.3406/ecoru.1996.4822.

Muller, P., 2000. La politique agricole française: l'État et les organisations professionnelles, *Economie Rurale*, 255, 33–39.

Ollivier, G., Magda, D., Mazé, A., Plumecocq, G., Lamine, C., 2018. Agroecological Transitions: What Can Sustainability Transition Frameworks

Teach Us? An Ontological and Empirical Analysis, *Ecology and Society*, 23, 2. https://doi.org/10.5751/ES-09952-230205.

Petersen, P., Mussoi, E. M., Dal Soglio, F., 2013. Institutionalization of the Agroecological Approach in Brazil: Advances and Challenges, *Agroecology and Sustainable Food Systems*, 37, 1, 103–14. https://doi.org/10.1080/10440046.2012.735632.

Petersen, P., Da Silveira, L. M., Bianconi, F. G., de Almeida, S. G., 2017. *Método de Análise Econômico-Ecológica de Agroecossistemas*, Rio de Janeiro, AS-PTA.

Rivera-Ferre, M. G., 2018. The Resignification Process of Agroecology: Competing Narratives from Governments, Civil Society and Intergovernmental Organizations, *Agroecology and Sustainable Food Systems*, 42, 6, 666–685. https://doi.org/10.1080/21683565.2018.1437498.

Schmitt, C. J., 2016. A transformação das "Ideias Agroecológicas" em instrumentos de políticas públicas: dinâmicas de contestação e institucionalização de novas ideias nas políticas para a agricultura familiar, *Política & Sociedade*, 15, 0, 16–48. https://doi.org/10.5007/2175-7984.2016v15nesp1p16.

Schmitt, C. J., Niederle, P., Ávila, M., et al., 2017. La experiencia brasileña de construcción de políticas públicas em favor de la Agroecología, in Sabourin, E., Patrouilleau, M. M., Le Coq, J. F., et al., *Políticas públicas a favor de la agroecología em América Latina y El Caribe*, Porto Alegre, Evangraf/Criação Humana, Red PP-AL, FAO.

Schmitt, C. J., Porto, S. I., Lopes, H. R., et al., 2020. *Redes de agroecologia para o desenvolvimento dos territórios: aprendizados do Programa Ecoforte*, Rio de Janeiro, Articulação Nacinal de Agroecologia – ANA.

Sjöblom, S., Godenhjelm S., 2009. Project Proliferation and Governance—Implications for Environmental Management, *Journal of Environmental Policy & Planning*, 11, 3, 169–85. https://doi.org/10.1080/15239080903033762.

Szwako, J., Lavalle, A. G., 2019. "Seing like a social movement": institucionalização simbólica e capacidades estatais cognitivas, *Novos Estudos CEBRAP*, 38, 2, 411–434. https://www.scielo.br/pdf/nec/v38n2/1980-5403-nec-38-02-411.pdf.

Viana, J. P., 2015. *Leveraging Public Programs with Socio-Economic and Development Objectives to Support Conservation and Restoration of*

Ecosystems: The Price-Support Policy for Socio-Biodiversity Derived Products and the Green Grant Programme of Brazil. Brasília, Institute for Applied Economic Research (IPEA). https://www.cbd.int/ecorestoration/doc/Brazil-case-study-Final-Version-20150114.pdf.

Teaching, training and learning for the agroecological transition: A French-Brazilian perspective

Moacir Darolt, Juliette Anglade, Pascale Moity-Maïzi, Claire Lamine, Florette Rengard, Vanessa Iceri, Amélie Genay, Cristian Celis

1. Introduction

During the so-called period of "agricultural modernization" (from the 1960s–1980s and beyond), in France as in Brazil and Latin America, teaching and training policies for farmers and agronomy professionals were oriented towards the diffusion of technologies from industrial agriculture, that responded to an economic rationality and to productionist models (Sarandon, 2016). Many ancestral practices developed by peasant and traditional communities were devalued and invisibilized (Sousa, 2017). The agricultural knowledge and innovation system became deeply anchored in a technological determinism linked with causal model of the processes and phenomenas (nomological laws) setting what should be learnt and taught, and then put in practice. Against these approaches and their negative effects, agroecological learning is considered as another way of thinking agriculture often described as pluri-epistemic, valuing different kinds of knowledge including experiential ones, and variations depending on where and how it takes place and on each individual choice.

In Brazil, agroecological training appeared in the end of the 1970s, within social movements, unions, associations, cooperatives, and non-governmental organizations (informal sector), if we refer to the classical distinction between formal and informal learning spaces or sectors (Anderson et al., 2019) associated with what was called "alternative agriculture" (Lamine et al., 2015). In the same time, stakeholders of public teaching and training establishments (formal sector), along with student groups, began to discuss alternative technologies for small farmers (Sousa, 2017). This process intensified in the 1990s and 2000s with the multiplication of networks dedicated to training farmers and advisors. Social movements also influenced this process of institutionalization of agroecological training (Niederle et al., 2019). This one took different forms, namely: teaching units in diverse disciplines (natural and social sciences) at different education levels (technical and higher education); groups connected with universities and public research centres to train students, farmers, technicians and citizens; research and diffusion projects. In Brazil, there are currently 230 state-recognized degree delivering agroecology courses and the number of formal agroecology courses grew by 70% between 2013 and 2017 (Engelmann and Floriani, 2018). Other short-term trainings and informal learning experiences in agroecology were stimulated by public policies such as the National Policy of Agroecology and Organic Agriculture (PNAPO) launched in 2012 and program such as Ecoforte, aimed at reinforcing agroecology networks (Sambuichi et al., 2017; see Lamine et al., in this book).

In France, we find a different context, as agroecology emerged later in politics (Lamine et al., 2019), while the *"ecologisation"* of agricultural policies had already taken other forms, with a fairly institutionalization (early 1980s) of organic agriculture anchored in a movement to (re)diversify agricultural models (Allaire, 2002). The first public education programs in organic agriculture were launched in 1985, but they were then limited to a few agricultural schools and only reached very concerned and convinced persons (Morin and Minaud, 2015). By the end of the 1980s, the Department of Education of the Ministry of Agriculture had created a network devoted to organic agriculture and to legitimize it in agricultural training courses. In 2012, a national agroecological policy was launched, seeking to englobe all agricultural sectors (whereas in Brazil, agroecology was oriented towards family agriculture and already since the early 2000s). This *"Plan Agroécologie pour la France"* was translated into a host of programs in which professional training is an important institutional objective, to renew the trainers' frames of reference for agricultural school trainers. A specific national plan devoted to agricultural teaching (*"Plan Enseigner à Produire Autrement"*) included agroecology into all the curricula and diploma, both in the "agricultural high-schools" and in the "professional centers."

Recent debates on agroecology training and education highlight gaps and unresolved questions in Brazil (Aguiar, 2017) and in France (Mayen, 2013; Chrétien, 2015). There is a consensus about the fact that training and education for the agroecological transition must be supported by new forms of pedagogy due to the triadic nature of agroecology (science, practice and social movement) (Stassart et al., 2012) which imposes multidisciplinarity and non-hegemonic posture or knowledge.

The agroecological transition also needs a post-normal approach to education in which affective and ethical dimensions should be (re-)valued and articulated to cognitive ones (Simmoneaux et al., 2016) in order to help students to deal with complex socio-scientific issues. Some academics inspired by an ecocentric philosophy and an ethic of care, move away from a human-centred worldview toward an Earth-centered one (Molina-Motos, 2019). Among these approaches, the concept of "eco-formation" (Moneyron, 2018) was suggested in order to promote robust ecologization processes (Silva et al., 2019). Other authors, in line with the libertarian philosophy of Freire, claim for a deeply political pedagogy in order to support a subversive process of social transformation for food sovereignty. The emerging current of "learning for transformation" advocates a critical learning approach that transgresses the binary divide between informal-formal education and suggest a collective strategy that transcends individuals to build capacity to understand inequalities and oppression within a theoretical perspective of positive change (Anderson et al., 2019). In that way agroecological education advocates a shift from a diffusionist, top-down view to participatory approaches that value farmers' knowledge and horizontal "dialogue of knowledge" (Silva et al., 2019). In practice, new modalities of training often draw on active pedagogies with, for example, a larger role accorded to experimentation (Mayen, 2013), and problem learning methods applied to real-life open-ended cases (Cuadra and Francis, 2014).

In this chapter, we investigate the relative place of determinist and open-ended perspectives in the agroecological transition (AET) courses and pedagogies. Our hypothesis is that teaching and training agroecology mobilize two dimensions in tension: a determinist conception of agriculture based on stabilized knowledge and a conception focused on progressive and necessary adjustments to risks and uncertainties valuing the formative nature of controversies (Prochasson and Rasmussen, 2007). These dimensions can enter in tension or in co-articulation. Two major questions guide this research: how does agricultural education historically anchored in a determinist perspective, integrate open-ended perspectives? What articulations, resistances or tensions do we find in different backgrounds?

We draw on six case studies: two cases concerning *teaching programs in public agricultural schools* (state centres for professional teaching in Brazil – CEEP, Paraná; and a professional training centre in Fra

hree cases in *alternative networks* (the Latino-American School of Agroecology – ELAA) in Paraná-Brazil, a farmers' education program developed by the Popular Education Institute (IEEP) in Paraná-Brazil and a mentorship program developed by the *InPACT* network (Initiatives for a Territorial and Citizen Agriculture) in Ardèche-France, and finally, a social experimentation recently launched in an INRA *public experimental station* in Mirecourt (Lorraine-France). Depending on the cases, our analyzes are built on documents, interviews with teachers, trainers, facilitators, and learners, ex-post program evaluation questionnaires, audio and video records, and ethnographic observation.

2. Case studies

2.1. Technical courses in agroecology in state centres of professional education (CEEP): Brazil

The state of Paraná, a pioneer in Brazil, started to put in place technical courses in agroecology in its agricultural secondary schools as early as 2004, in the regions where family agriculture and diversified production were dominating. The participants are principally youth aged from 15 to 24 years who already work within the family farm (Gomes, 2014). The programs are validated within the framework of regional seminars organized by the state department of education and in the municipalities where the secondary schools are located, with the participation of the local community (parents, farmers, students, teachers, social organizations), so as to adapt the program to territorial realities. The agricultural model defended by the CEEP is that of a family agriculture that respects local ecosystems and agroecology constitutes a significant facet of training programs. The general goal is to train critical professionals able to implement agroecological transition processes (SEED, 2017). These secondary schools train students on farms so they can conduct experiments related to management of agroecological systems. The schools also involve territorial actors to foster the insertion and installation of youth in their area (municipal agricultural secretaries, research and extension institutions, farmers' associations, banks).

Around half the hours of the training (that lasts three years full-time) is dedicated to technical disciplines with a clear focus on agroecology, and the other half to non-technical disciplines among which some (philosophy, sociology and the arts) are oriented towards deepening humanist training, which seeks to shape protagonist-oriented students involved in social issues. The courses are organized in alternating phases (known as the *pedagogia da alternância*), with three weeks of courses at school and one week in the family farm or one of another rural community with ties to the training. This aims at the intrinsic articulation between formal education and the specific reality of the rural populations (Molina and Sá, 2012). The organization of the training in alternating phases requires a reorganization of the pedagogical process, a flexibility to adapt to the students' projects and a permanent accompaniment, carried out by teacher-supervisors. The testimony of a teacher reflects this aspect: *"the school calendar needs to give more time for structuring the life plans of the students and their families, to better understand the local reality and their tacit knowledge."*

The principle of alternating phases is particularly adapted to the socio-economic situation of the students, who often come from low-income peasant families, facing many constraints and uncertainties. It allows them to maintain a presence on the farm while undergoing training. The goal is to encourage students to develop tangible projects and activities on the family farm and within their communities of origin. The program thus encourages the youths to remain in the rural setting, creates alternative work and revenue options and confronts the students with concrete and situated challenges related to the agroecological transition and its various scales (family farm, community, territory). It prepares youth for becoming autonomous, both socially and economically, and helps them in defining their future: *"I thought that it also could help me become independent, and do something with my life,"* explains one student.

This case shows two aspects of education in agroecology that stems respectively from a determinist and an open-ended perspective of the AET, which appear as complementary. On one hand, the teaching in farm-schools provides students with controlled experimental plots, which relates to a determinist view of the transition (targeted goals, predefined steps). On the other hand, the conception of the program, oriented towards a whole life project with experiences on the family farm or in the community, and possible adjustments to students' plans in order to adapt to their specific contexts, relates to an open-ended perspective.

2.2. The public agricultural teaching programs in France

The Professional Agricultural Baccalaureate and the Professional Certificate in Farm Management (BPREA) are diplomas that allow one to obtain a status of "professional agricultural capacity." These vocational qualifications are required for an access to public subventions for farm establishment. The former is provided by an upper-level Agricultural Secondary School in a three-year program and the latter in a professional training centre (CFPPA) for adults. In this case, the program consists of 10 months of training with the elaboration of a project for establishing a farm (that must be validated by a jury), and periods of alternating internships, that last 4–10 weeks with one or more farmers. The pedagogical content, renewed in 2012 in order to integrate agroecology, is defined by the State due to it being a public diploma based on general skills specified at the national level. Social sciences represent a much lower part of the program than in the above Brazilian case. The nature and format of teaching were fairly thoroughly revisited with the new curricular framework. Collective activities on local farms are now favoured which allows students not only to access to know-hows but also to confront them collectively within real world situations. This pedagogy is designed to generate stronger discussions and reflexivity. The new curriculum thus also includes social questions, for example controversies related to animal welfare, water quality, the elimination of glyphosate.

For adult training, the new curriculum is based on "explicative interviews" with established farmers, seeking to list and analyze all the functions and activities they undertake. These interviews lead to the establishment of collective "Descriptive Activity Plans" that should be subsequently adapted depending on the region and contexts. Teachers are then expected to establish "Significant Professional Situations" where each one can test different kinds of knowledge in action. These new teaching methods reflect a more reflexive vision of the AET relating to concrete farm projects. Teachers see in these new pedagogies ways to discover and confront agricultural models and to change their teaching habits and contents. In the framework of these transformations some research programs aim to identify and formalize pedagogical innovations and new agricultural practices emerging from the diversity of "Significant Professional Situations." This was the case in the AP3A project, which sought to accompany students and teachers towards an AET with perennial crops, in partnership with institutions dedicated to research, development and training (Genay, 2018).

We studied the specific space provided for experimentation in the training syllabus. In the framework of the AP3A project, a diversified orchard plot was created on the farm-school to allow the definition of a succession of pedagogical sequences based on plot conception, plot objectives (zero chemical inputs, economic viability, functional biodiversity, optimization of time management) and plot creation, with students. Our interviews (with a dozen students in collective sessions) sought to analyze how these students were involved in this experiment, what they learned, and how they could use it in their own farm projects (Celis, 2018).

For teachers the primary goal was first and foremost the learning of innovative practices. Our analysis shows that the knowledge circulating in this experimental local network primarily relates to crop practices and management and to general principles of agroecology. But it also shows that the learning processes are influenced and reinforced by the collective dimension of learning situations characterized by discussion, debates relating to experiments and practice reflexivity. The students were strongly involved in the process but felt destabilized by the uncertainties linked to the very principle of learning "along the way": it is built through work situations that are always singular, it is marked by unexpected results or parameters, it is regularly rebuilt in a dialogue between trainers and learners (Metral et al., 2016). Finally, our analysis shows that the processes of knowledge appropriation – that makes it actionable – partly depends on the experiential background and expectations of students. In sum, if the process of conception for this experimental plot was anchored in a determinist vision of the AET (with pre-determined objectives and steps), our analysis highlights that it is the students critical learning processes that allows this experiment to be adapted locally in a more "open-ended" perspective. Indeed, the participative and collective dimension of the experiment generated unexpected outcomes.

2.3. Latin-American school of agroecology: ELAA (Lapa-Paraná-Brazil)

Since 2002 the Landless Workers Movement (MST) of Paraná initiated actions seeking to fortify their pedagogy in agroecology. The establishment of technical courses in agroecology beginning in 2004 are due to pressures from social movements and took place within a wider national movement in favour of rural education (*educação no campo*) (Guhur et al., 2016). The ELAA was the first Brazilian experience in agroecology at the university level, inaugurated in 2005 with the goal of establishing a technical qualification and training young educators and Latin-American "activist-technician-educators" to work in their countries of origin. The school provides students with certification as "Technicians in Agroecology" and as rural professors with a Diploma in Rural Education, Natural Sciences and Agroecology. As of this date, the ELAA has trained about 180 students. The training takes place in a residential campus over three and a half years.

The training programs adapt to the daily-life preoccupations of farmers and social movements, as well as to the reality of production systems of each region where the schools/centres are located. Because of this, a portion of the program is unvarying while another portion seeks to respond to practical questions related to regional food systems (Caldart, 2016). The training is based on international experiences of Vía Campesina Agroecological Institutes, demonstrating the "Campesino a Campesino (CaC)" method and the *"Diálogo de Saberes"* (Rosset and Barbosa, 2019). The construction of the programs associated with these social movements (MST and Vía Campesina) involves professors from public universities (Federal Institut of Paraná – IFPR) conforming to the national rural education policy (*Educação no Campo*). The course is financed by resources of the National Education Program for Agrarian Reform (PRONERA). The implication of actors/stakeholders relies on the permanent team (coordination and public direction of schools composed of social movement representatives, teachers of the university) and the mobilization of professionals (internship tutors, partner businesses and local actors).

The curriculum is based on pedagogies inspired by popular education (Paulo Freire), by the « *Educação no Campo* » movement and by the historical dialectical materialism (Brandão, 2014). Concretely, the courses of "Technicians in Agroecology" are composed of six parts, with

60% representing general or humanist disciplines and 40% technical aspects. The notion of *work* is a fundamental element, at the crossroad of the different disciplines. The school-community relationship is seen as a strategic element as well as the political training (Guhur et al., 2016).

The training with alternating periods brings together two steps: the *Tempo-Escola* (40–75 days) during which the training activities are organized intensively, the students live together in the school full-time; and the *Tempo-Comunidade* (around 60–90 days) during which the students return to their communities of origin. The aim is to provide a complementarity between practical and theoretical knowledge, based on guided studies, research, workshops, internships, systematization of agroecological studies, dialogue of knowledge and *praxis* in the sense of Paulo Freire (Freire and Nogueira, 1993). At the end of the course, the students develop a final thesis associating research in agroecology with the *Diálogo de Saberes*, which is considered as the most adapted method for addressing agroecological complexity and guiding relations between technicians and peasants. This thesis is supposed to retrace the family history, characterize the system of production, address the necessary changes in the agroecosystem (Guhur et al., 2016).

This case illustrates an open-ended perspective of the AET, with flexible and multidimensional pedagogical models, with education principles based on dialogue rather than on the transfer of knowledge, focused on practical reality or experiences, and seeking for a balance between technical, social and political training. It also illustrates the ambition to train students and peasants as researchers of their agro-ecosystems, through dialogue of situated knowledges.

2.4. Faxinal Emboque, São Mateus do Sul, Paraná: Brazil – *The institute of popular education* – Instituto Equipe de Educação Popular **(IEEP)**

This experience relates to non-formal education of farmers in transition towards agroecology in a traditional[1] Brazilian community (Faxinal Emboque) that developed a series of agro-sylvo-pastoral actions

---

[1] « Traditional communities » refer to the set of social groups that defend their respective territories and their permanence in it, seeking cultural autonomy and adaptive practices. This concept was incorporated into the Brazilian Constitution of 1989 and the Law of the National System of Conservation Units.

in environmental protection zones. The Faxinais-type communities combine the use of common resources (land use, pasture and forest resources) for free range livestock with crops (self-consumption and surplus commercialization), as well as the preservation of forests (transfer of management and conservation to local populations). If on one hand the traditional Faxinais model (agro-sylvo-pastoral) is legally recognized by the state since 2007, on the other this model has difficulty confronting insufficient income levels in these communities. Furthermore famers suffer from a lack of knowledge and technical accompaniment to change their systems. To tackle these challenges, producers organized themselves into networks in order to raise their political demands and develop collective actions, such as the *Terra Faxinalense* project. It is an initiative carried by members of the Emboque community and the Institute of Popular Education (IEEP) in order to not only reinforce agroecological practices but also the traditional way of life, based on the use of commons (Iceri and Lardon, 2018). The IEEP was created in 1995 to promote sustainable local development, based on the principles of agroecology with methodological principles of solidary and autonomous networks of collaboration, and works with rural organizations (rural extension, accompaniment, technical assistance) and scientists. The local interest for curbing agroecology is tied to the economic impact of the reduction of chemical fertilizers and conventional seed purchases as well as the increased power of certain producers, candidates or elected officials in local public offices. These conflictual relations are linked to the empowerment processes that were generated by the IEPP action. The popular education practiced by the IEEP is based on a learning process driven by actions, exchanges between stakeholders and networks and trials carried out on their farms.

The learning by *exchange* between individuals and collectives is possible through farm visits, meetings, workshops and discussion in WhatsApp groups. The exchange and sharing of knowledge is also sustained by the establishment of networks with universities, researchers and accompaniment institutions, which requires strong facilitation skills. Finally the *"trial"* dimension is present at the individual as well as collective levels. In the case of Faxinal Emboque, the production of *gabiroba* (a local plant) ice cream in partnership with a local ice cream producer is part of an initiative carried out by a woman, farmer in the community, while experimentation with vegan recipes, change of pork breeds and production of preserved pork meat exemplify collective innovations and learning experiences.

This case shows that the practice of agroecology in local traditional communities questions the anthropocentric relations between society and environment and puts forwards other elements (animals, such as pigs and fences, are symbolic in the Faxinal system). It also questions the paradigms of technological progress and linear development, and the patriarchy, scientific and institutionalized knowledge seen as domination norms. The agricultural practices in the Faxinal system embody a process of agroecological transition in the making.

2.5. Mentorship system (InPACT collective): Ardèche – France

In France, more than a third of farm projects are set up on farms that were not inherited from parents or grandparents. The profiles of the individual carrying these projects are heterogeneous with certain traits in common: the search for an attractive quality of life (relating to social, environmental and food factors), a strong blending of professional and personal goals, the ambition to become part of a community of practice (Van Dam et al., 2009), and, concretely, a high portion of projects aiming for organic production. Yet, the framework mobilized to support farm set up projects are not adapted to these profiles. It was thus necessary to invent another approach. It is what the network called InPACT in Ardèche did by establishing a system of mentorship, based on individual accompaniment of future or new farmers through apprenticeship with experienced peers. Over a period of about 6 months, associations (members of InPACT) met four times in a steering committee made up of their employees, farmers and researchers, in order to co-build this system. Indeed, this mentorship system already exists, for example in agricultural test-spaces (Fabre et al., 2016) where aspiring farmers are hosted on a collective farm for approximately two years[2] (Bui et al., 2016). Subsequently, an analysis of needs and first experiences of already existing mentorships was co-built. Concretely, the mentorship involves a series of periods where diverse activities are carried out in pairs on the mentor's or aspiring farmer's farm. The transmission of professional knowledge as

---

[2] https://reneta.fr/Le-test-d-activite-agricole, accessed 15/6/2019.

both content and process (Chrétien, 2015) is carried out at one place of work and *through* work, out from an educational centre.

The mentorship is based on the learning of the farm trade as a "life trade," in which work, life goals, social participation in the community and professional engagements are articulated together (Chrétien, 2015). A co-built evaluation of the future farmer's needs is realized through a dialogue with the mentor and with an accompanying organization in order to define adapted modalities (frequency, place, activities). The mentorship role is based on the skills and the stage of the future farmer's project. Three profiles of them have been identified: emerging, in realization (less than 6 months), young established (less than two years). The shared activities also depend on the mentor's specific skills and on his available time. The individual mentorship can be combined with "*à la carte*" collective training. For example, an "emerging" project would especially require general technical knowledge (plant, irrigate, diagnose diseases, valuing the production etc.). A project "in realization" would require more specific technical knowledge, management and accounting ones, but also more organizational aspects tied to time management and to the balance between social and work life. Beyond experiential knowledge and "real stuff," the mentor is also a facilitator to link the future farmer with different networks (other farmers of the region, networks of transformation and commercialization of products, local stakeholders in general) and with key values and unformal rules such as mutual help.

In terms of learning modalities, our analysis shows that mentoring encourages learning through "seeing," "doing" and "living" (all part of an experiential learning). Through seeing: observation, exchange and advice – for example the aspiring farmer would accompany his mentor who explains him/her the tasks that are carried out, allowing a progressive learning of steps and technical knowledge, in the mentor's farm. Through doing: the aspiring farmer works with his/her mentor most often on his farm, in repeated cycles which allows the first one to become familiar with technical gestures (know-how). Through living: the immersion of the aspiring farmer in the mentor's farm allows him/her to acquire work habits and know-how related to the farm trade confronting decision making, adapting to mishaps, reorienting their project, organizing their day, exchanging about the work/life balance. These modalities are themselves adapted or modified during the process of the mentorship. This system is clearly anchored in an experiential and interactive perspective of the AET. First, the objectives and steps of the farm set up as well as the aspiring farmer's professional path are redefined along the way and then the imperative of adapting to local contexts is translated by a double contextualization articulating the situation of the mentor and the future situation of the aspiring farmer. Innovations and conversion to AET emerge from interactions and knowledge is a result of situated working collaborations.

2.6. Reinventing experimental farms as new learning spaces: INRA Mirecourt, Lorraine – France

From 2016 to 2019, the INRA experimental station in Mirecourt carried out a social experimentation that led to conceive new modalities to share knowledge and experiences about organic diversified crop/livestock systems (Anglade et al., 2018, 2019). The ambition was emancipatory as it aimed at accompanying the transformation of farming practices towards sustainable local agri-food systems. 60 farm visits were organized in order to answer the increasing demand from teachers and advisors – linked to the recently set up new agroecological national policy – and gathered more than 1200 participants from the agricultural sector (27% farmers, 57% students in technical and higher education in agriculture, 16% from technical and research institutes).

Mirecourt experimental station was historically – like most other stations – devoted to the production of empirico-formal knowledge in controlled conditions (reductionist approach) in order to optimize agronomic or zootechnical performances. This led to produce prescriptions (e.g. feeding rations) that were disseminated by the advisory system and rare short thematic visits. From 2004 the conversion of the whole station (240 ha) towards organic farming gave a new epistemic orientation to the experimentation that became systemic and generative. A pragmatist approach based on step-by-step design was adopted (Coquil et al., 2014) with an overall aim of both material and decisional autonomy. The station is not anymore seen as a place for testing and verifying pre-defined questions but as a starting point for establishing and resolve new questions raised by the local human and non-human environment. This new regime of knowledge production embodies an open-ended perspective to AET, with no pre-defined decision rules, integration of the intrinsic complexity of the systems and adaptation along the way. This also means that the management rules have a low degree of stabilization and ability to be transferred to other contexts. The learning framework that was designed combines four active learning approaches: phenomenon-based, inquiry-based, experiential learning, and social learning. Based on a roles' inversion and on a collective inquiry approach, visitors are invited to portrait the farm based on cognitive maps, drawings, poetry, land art etc., bringing together perceptions, analyzes, experience, intuition and imagination. Teachers and the stations' practitioners accompany the visit as inquiry-facilitators, trying to provide appropriate resources when needed through testimonies (storytelling). The pedagogic sequences always take place in the working place and begin in silence with a sensuous immersion and a careful observation. This step aims to facilitate the connection to a sentient reality and its phenomena. Then it articulates several phases of individual appropriation and of interactions among visitors and with the station's staff allowing discussions over gaps and differences between each one's observations and "negotiations" of meanings based on argumentation processes.

The evaluation of the trainings was not standardized and not driven by content but learner-centered. It was carried out by the participants themselves (auto-evaluation) on the basis of a virgin reflexive logbook to record astonishments and feelings all over the visit, with ex-post forms made of open-ended questions action-oriented. A variety of learning trajectories were revealed, ranging from simple single technical adoption to systemic transformations impacting belief systems. Among the station's staff, collective debriefings were done after each visit (during two years) to share progressive learnings and concerns about this new teaching experience that requires to listen and to gain confidence in the value of experiential knowledge (compared with academic knowledge). *"By starting from learner's questions and not anymore from prepared speeches and figures, the most difficult for me was to often say I don't know."* For some students, it was a breach of didactic contract: *"If even the researchers don't have answers, who will tell what and how we should do?"* In this case, it is both the experimental station itself and the modes of learning that shift from a determinist perspective to reach an open-ended perspective based on step-by-step and collective experimentation and learner-centered processes integrating ethics and affects. This shift generated strong epistemic and human tensions. Some teachers and part of the station's staff expressed feelings of resistance and fear in front of the higher degree of uncertainty, and some nostalgia for the positivist rationality which allowed them to guide action based on well-known solutions, understood as sure and "true" knowledge.

3. Discussion: The modes of articulation between determinist and open-ended perspectives

Our transversal analysis of Brazilian and French case studies shows how programs and education contents in agroecology can be conceived and organized in a variety of processes. It also reveals the tensions but also the possible articulation of determinist and open-ended perspectives. Both of them are mobilized in the different case studies we described, with contrasting combinations. Tab. 1 presents an overview of the two perspectives based on our six cases.

The open-ended perspective is more present in training and teaching programs of alternative networks in Brazil (*ELAA* and *Faxinal*) and France (Inpact). The determinist perspective is more present in public education programs, which have clearly targeted objectives and steps, both in the French (*BPREA*) and Brazilian case studies (agricultural secondary schools), nonetheless with varying articulations with an open-ended perspective, namely through an alternating training model (between farm and classroom) allowing concrete and situated interactions on the farm. It is increasingly acknowledged that training and teaching in agroecology require programs and content which stimulate acquisition of "autonomous and critical thinking" supported by social sciences approaches. This goes hand in hand with an adaptive perspective. This "openness" is is closely related to the way programs are framed. Public trainings are conceived in a centralized manner through national/regional curricula, in contrast to alternative network programs established from local demands. On the one hand, the recent revision of these curricula tied to agroecological policies have led to supporting these programs with more locally anchored situations (such as the BPREA in France). On the other hand, for the Brazilian case study, the programs are built in partnership with civil society organizations in the regions involved. The MST proposals and works reveal its ability to implement new pedagogical programs inside the formal schooling system(s), even under contradictory and often hostile conditions, reflecting a "learning for transformation" stance (Anderson et al., 2019).

In their teaching modalities several courses use immersion in concrete situations, thus developing what some call "open-ended case studies" (Francis et al., 2009) or "open ended case learning methods" (Cuadra and Francis, 2014). Some cases reveal learning methods based on exploratory and creative inquiry that evolves continuously between openness

Tab. 1: Comparison between determinist and open-ended agroecology teaching and training programs in Brazil and France.

	Determinist (D)	Open-ended (OE)
Objectives, steps	Clearly targeted, pre-defined (at a national/regional scale)	Adaptive and experimental, anchored on a territorial scale
Construction of programs and content	Centralized (e.g. national public curricula)	Decentralized, participative and flexible
Teaching modalities	"top-down": demonstration, experimentation, farm schools, classroom	"bottom-up" and negociated: participatory, without hierarchy, based on field experience
Learning methods and types of knowledge	Learning by experiments in controlled conditions. Stabilized and analytical (expert) knowledge transmitted in school (out of farms)	Experiential knowledge, *Diálogo de Saberes*, active and collective learning. Immersion into concrete cases, alternating school and farming works
Stance of trainers/teachers	Diffusionist approach	Socio-constructivist approach
Evaluation methods	External, a posteriori (verification of learning objectives and criteria fixed a priori by experts). Evaluation centered on performances	Learner-centred, interactive and continuous forms of evaluation, including social parameters (such as work)
Visions of agroecology	Unidimensional (focused on techniques)/individual focus	Multidimensional and trans-disciplinary/collective focus/care approach of human and non human
Agroecological transition/transformation	Transition from one model to another	Experimental learning for technical, social and political change

Source: authors

and commitment, mobilizing abductive reasoning. It relies on a "multilogue," linking individuals, groups and non-humans that constantly reopen the learning trajectories. These learning approaches also encourage a critical awaking and prepare to be life-long learners in order to deal with complex, uncertain and changing contexts. In this sense, pedagogical models for AET require concerted dialogue between community leaders, educators, students, and their families about how to build new agricultural systems upon a foundation of existing cultural traditions

(Meek, 2016). The open-ended perspective also takes into account students' projects, both personal and professional, as well as their evolutionary contexts and projects. In the six cases, we also highlighted that questions about « work » are central and gateway for an agroecological learning process, but source of tension between production objectives and projects enhancing quality and care as essential life values. We also see that the substantial part dedicated to non-technical disciplines, oriented toward deepening humanist training in agricultural secondary schools in Brazil, has no equivalent in France where this human dimension is mostly supported by the mentorship system, out of school.

The articulation of determinist and open-ended action finds its footing with the implementation of an alternating pedagogy (*pedagogia da alternância*) which is the principal modality of agroecological teaching in Brazil, connected with the paradigm of *Educação no Campo* (Caldart, 2016). On one side, lessons are part of a curriculum that has set aims, steps and learning objectives. On the other, a training alternating periods between classroom and field allows each student to build his/her own project linking it with real experiences on farm, taking into account the family's background as well as constraints and necessary adjustments. Nonetheless, this back-and-forth pedagogy requires conditions that are difficult to establish (flexibility of programs, cost of trips), especially in the post-2016 period, as a result of political changes in the country threatening those forms of teaching and learning.

More and more teachers and trainers strive to shift from a diffusionist approach toward a socio-constructivist posture. They recognize the fact that knowledge is built through social interaction relied to "real life" and situations. But this shifting posture can generate epistemic and human tensions that express feelings of uncertainty. These more open-ended perspectives also open the black-box of modalities and criteria of learning evaluation, concerning both learners and teachers. Knowledge is no longer just external and related to technical objectives established ahead of time, but also built during the learning process through interactions between learners and experts taking account dimensions of human actions on farm. New criteria and forms of evaluation for learners and trainers must be invented, taking into account, inter alia, self-actualization and socio-cognitive moves.

The two perspectives (D/OE) we described reflect contrasting visions of agroecology (Tab. 1) that can generate conflicts within pedagogical teams. Our case studies suggest an evolution from an individual to a collective focus, in tune with authors who show that collective learning processes are acts of resistance against the individualizing tendencies. They also highlight that informal and collective learning spaces, embedded in communities, networks and social movements, are more attractive and viable (Anderson et al., 2019). In Brazil the majority of public trainers teach agroecology as a counterpoint to the paradigm of modern capitalistic agriculture, placing value on traditional peasant knowledge. They defend a systemic approach of the production process. As for teaching programs in alternative networks, they see agroecology as a socio-political movement fighting for the construction of a new cognitive and technical paradigm adapted to family agriculture and taking into account various dimensions (non-human and human relationships, health, care, energy, aesthetic, leisure, solidarity, quality of life and food). Regarding agricultural practices, as we see in the case of *Faxinal*, a reference to agroecology allows for a requalification of some traditional know-how (traditional extractivism, agroforestry, sylvopastoral practices) and associated products. In France the traditional empirical knowledge has long been devalued and invisibilized in favour of analytical and expert knowledge. The challenge is now to (re)build and legitimize other kinds of knowledge, that is, more experiential, situated and sensitive, and to develop new forms of nature-human relations. In the cases studied – both in Brazil and France – open-ended perspectives to AET lead to favour learning for broad social and political transformation, while more determinist perspectives enhance technical-productive and economic dimensions.

4. Conclusion

The six cases show a tendency of evolution from a determinist perspective to agroecological training and teaching to an open-ended one, particularly over the past ten years, both in Brazil and France. In all cases, we find articulations between both perspectives, but in some cases, especially in alternative networks, actors assert more explicitly an open-ended perspective, reflecting a conception of agroecology as multidimensional, with individual farm goals inseparable from collective life goals. The open-ended perspective explicitly recognizes conflict as a possible way out, or even as a learning method (conflict as a driver of change). Debate and conflict appear in all cases as reflecting resistances but also express a real capacity for critical position, in front of uncertainty. We also emphasize that tensions and compromises are intrinsically part of the transition of teaching. Nonetheless, the didactic modalities that allow for the establishment of a pluri-epistemological dialog, with a true emancipatory and transformative ambition (learning for transformation) remain to be explored. Experimental farms in both countries could become privileged places for this exploration. Our confrontation of Brazilian and French case studies provides comparative insights, source of inspiration on both ends. The French experience of a mentorship network could be adapted to Brazil, in particular in areas where family farmers have little public support but are collectively organized. On the other hand, the *campesino-a-campesino*, *diálogo de saberes* and *educação no campo* methods of Brazilian social movements could serve, and in some cases already served, as inspiration for French networks. But what strongly characterizes Brazilian cases is the goal of developing the critical capacity of students (regarding their own situations and the structural inequalities of their society), tied to the principles of popular education that also have an increasing influence on French alternative networks.

The choice for teaching and training in an open-ended agroecological perspective encourages and promotes a change in pedagogy and learning postures with a greater emphasis on (i) the participation of local actors to build programs and contents and curricula, (ii) flexible pedagogical models that take into account technical, economic, social and sensitive aspects of students' lives and contexts, (iii) innovative learning methods emerging also from collective exchange and experimentations, (iv) values relating to dialogue, reciprocity, solidarity and horizontality between actors engaged towards agroecological transition, and (v) collective engagement and critical reflexivity from both teachers and learners, as necessary to change agricultural practices as parts of a larger technical and socio-political project.

Acknowledgments

We would like to thank the two anonymous reviewers for their useful feedback. Thanks also to Morgan Jenatton for English-language editing.

References

Aguiar, M. V. A., 2017. A experiência de Educação como caminho para a construção da Agroecologia – Pontos para o debate, *Cadernos de Agroecologia*, 12, 1, 1–22.

Allaire, G., 2002. L'économie de la qualité, en ses secteurs, ses territoires et ses mythes, *Géographie, Economie et Société*, 4, 2, 155–180.

Anderson, C. R., Binimelis, R., Pimbert, M. P., et al., 2019. *Agriculture and Human Values*, 36, 521–529. URL: https://doi.org/10.1007/s10460-019-09941-2.

Anglade, J., Godfroy, M., Coquil, X., 2018. A Device for Sharing Knowledge and Experiences on Experimental Farm Station to Sustain the Agroecological Transition, in *13th European IFSA Symposium*, 1–5 July, Chania (Greece).

Anglade, J., Godfroy, M., Coquil, X., 2019. Awakening Senses and Sensitivity to Make Sense in Learning Approaches of Agricultural Sustainability, in *ESEE2019 – 24th European Seminar on Extension and Education*, 18–21 June, Acireale, Italia.

Brandão, J. D., 2014. *Educação e integração: A práxis educativa da Escola Latino Americana de Agroecologia – ELAA e da Universidade Federal da Integração Latino-Americana – UNILA*. Trabalho de Conclusão de Curso, Universidade Federal da Integração Latino-Americana, Foz do Iguaçu.

Bui, S., Cardona, A., Lamine, C., Cerf, M., 2016. Sustainability Transitions: Insights on Processes of Niche-Regime Interaction and Regime Reconfiguration in Agri-Food Systems, *Journal of Rural Studies*, 48, 92–103.

Caldart, R. S., 2016. *Escolas do Campo e Agroecologia: uma agenda de trabalho com a vida e pela vida!* Access on 11 October 2020. https://www5.unioeste.br/portalunioeste/arq/files/GEFHEMP/01_-_Escolas_do_Campo_e_Agroecologia.pdf).

Celis, C., 2018. Les processus de réappropriation des connaissances à partir des lieux d'expérimentation en Agroécologie dans un département de la région de Rhône-Alpes, *L'Étude de cas dans deux experimentations dans le milieu agricole de la Drome*, rapport de stage, Université de Strasbourg/INRAE/EPLEFPA, Valence, 49 p.

Chrétien, F., 2013. Les conceptions de la nature et du vivant : quelles places ont-elles dans les espaces d'apprentissage agricole ? *Pour*, 219, 3, 131–140.

Chrétien, F., 2015. Agriculteurs et apprenants au travail. *La transmission professionnelle dans les exploitations agrobiologiques: une approche par les configurations sociales et les situations d'interaction*, Thèse, University of Bourgogne.

Coquil, X., Anglade, J., Barataud, F., Brunet, L., Durpoix, A., Godfroy, M., 2019. TEASER-lab : concevoir un territoire pour une alimentation saine, localisée et créatrice d'emplois à partir de la polyculture – polyélevage autonome et économe. La diversification des productions sur le dispositif expérimental ASTER-Mirecourt, *Innovations Agronomiques*, 72, 61–75.

Coquil, X., Fiorelli, J.-L., Blouet, A., Mignolet, C., 2014. Experiencing Organic Mixed Crop Dairy Systems: A Step-by-Step Design Centred on a Long-Term Experiment, in Bellon, S., Penvern, S. (Eds.), *Organic Farming, Prototype for Sustainable Agricultures*, London, Springer, 201–217.

Cuadra, M., Francis, C., 2014. Experiential Learning Using Open-Ended Case Studies, in Eksvärd, K., Lönngren, G., Cuadra, M., Francis, C., Johansson, B., Namanji, S., Rydberg, T., Ssekyewa, C., Gissén, C., Salomonsson, L. (Eds.), *Agroecology in Practice: Walking the Talk*, org. Swedish University of Agricultural Sciences, Uppsala, Sweden: Reports Department of Urban and Rural Development, No. 1/2014, 41–55.

Engelmann, S. A., Floriani, N., 2018. Expansão da educação agroecológica formal no Brasil: construindo novas territorialidades nos últimos 17 anos, *Terr Plural*, 12, 1, 22–40.

Fabre, C., Moity-Maïzi, P., Cavalier, J.-B., 2016. Les espaces-test agricoles: expérimenter l'agriculture avant de s'installer, *Analyse* n°92, *Les publications du Service et de la Prospective*. URL: http://agreste.agriculture.gouv.fr/publications/analyse/.

Francis, C., King, J., Lieblein, G., Breland, T. A., Salomonsson, L., Sriskandarajah, N., Porter, P., Wiedenhoeft, M., 2009. Open-Ended Cases in Agroecology: Farming and Food Systems in the Nordic Region and the US Midwest, *The Journal of Agricultural Education and Extension*, 15, 4, 385–400.

Freire, P., 1992. *Extensão ou comunicação?* 10^a ed. Rio de Janeiro, Paz e Terra, 93.

Freire, P., Nogueira, A., 1993. *Que fazer: teoria e prática em educação popular*, 4a ed. Petrópolis, Vozes, 68.

Genay, A., 2018. Les apprenants du Valentin implantent un verger « zéro phyto », *Pour*, 233, 1, 13–19.

Gomes, D. A., 2014. Política pública de formação profissional do campo: Pedagogia da alternância, *Dissertação Mestrado*, Pontifícia Universidade Católica do Paraná – PUC, Curitiba, Brasil.

Guhur, D. M. P., Lima, A. C., Toná, N., Tardin, J. M., Madureira, J. C., 2016. As práticas educativas de formação em Agroecologia da Via Campesina no Paraná, *Cadernos de Agroecologia*, 11, 1, 21.

Lamine, C., Sibylle, S., Ollivier, G., 2015. Pour une approche systémique et pragmatique de la transition écologique des systèmes agri-alimentaires, *Cahiers de recherche sociologique*, 58, 95–117.

Lamine, C., Niederle, P., Ollivier, G., 2019. Dossier : Perspectives franco-brésiliennes autour de l'agroécologie – Alliances et controverses dans la mise en politique de l'agroécologie au Brésil et en France, *Natures Sciences Sociétés*, 27, 1, 6–19.

Mayen, P., 2013. Apprendre à produire autrement : quelques conséquences pour former à produire autrement, *Pour*, 219, 3, 247–270.

Meek, D., 2016. The Cultural Politics of the Agroecological Transition, *Agriculture and Human Values*, 33, 2, 275–290.

Métral, J.-F., Olry, P., David, M., Chrétien, F., Prévost, P., Cancian, N., Frère, N., Simonneaux, L., 2016. Ruptures ou ajustements provoqués entre pratiques agricoles et enseignement de ces pratiques, *Formation emploi*, Juillet-Septembre. URL: http://journals.openedition.org/formationemploi/4856.

Molina, M. C., Sá, L. M., 2012. Licenciatura em Educação do Campo, in Caldart, R. S., Pereira, I. B., Alentejano, P., Frigotto, G. (Eds.), *Dicionário de Educação do Campo*, Rio de Janeiro, São Paulo, Escola Politécnica de Saúde Joaquim Venâncio, Expressão Popular, 468–474.

Molina-Motos, D., 2019. Ecophilosophical Principles for an Ecocentric Environmental Education, Education Science, 9, 37. doi: https://doi.org/10.3390/educsci9010037.

Moneyron, A., 2018. *Agroécologie: quelle écoformation?* Editions L'Harmattan, Paris.

Morin, J., Minaud, B., 2015. L'agriculture biologique dans l'enseignement agricole. Panorama, freins et leviers, *Pour*, 227, 3, 207–215.

Niederle, P. A., Sabourin, E., Schmitt, C. J., Ávila, M. L., Petersen, P. F., Assis, W. S., 2019. A trajetória brasileira de construção de políticas públicas para a agroecologia, *Redes*, 24, 1, 270–291.

Prochasson, C., Rasmussen, A., 2007. Du bon usage de la dispute. Introduction, Mil neuf cent, *Revue d'histoire intellectuelle*, 25, 5–12.

Rosset, P. M., Barbosa, L. P., 2019. Territorialização da agroecologia na Via Campesina, *Agro Ecologia*, julho, 46–52.

Sambuichi, R. H., Moura, I. F., Mattos, L. M., Ávila, M. L., Spínola, P. A. C., 2017. *A Política Nacional de Agroecologia e Produção Orgânica no Brasil: Uma trajetória de luta pelo desenvolvimento rural sustentável*, IPEA Brasília, Brasil.

Sarandón, S. J., 2016. Potencialidades, limitaciones y desafíos para la introducción de la agroecología en la educación agrícola superior en la Argentina. El caso de la cátedra de agroecología en la Universidad Nacional de la Plata: Una experiencia de 16 años, *Agroecología*, 11, 1, 47–61.

SEED – Secretaria de Estado da Educação, Paraná – 2017. *Reestruturação do Plano de Curso Técnico em Agroecologia*, URL: http ://www.cee.pr.gov.br/arquivos/File/pdf/Pareceres_2017/CEMEP/pa_cemep_70_17.pdf.

Silva, J. C. B. V., Lamine, C., Brandenburg, A., 2019. The Place of Ecolearning in Ecologisation Processes in Family Agriculture in Paraná (Brazil), *Natures Sciences Sociétés*, 27, 1, 39–52.

Simonneaux, J., Simonneaux, L., Cancian, N., 2016. QSV Agroenvironnementales et changements de société: Transition éducative pour une transition de société via la transition agroécologique, *Diversité Recherches et Terrains*, 8. URL: http://dx.doi.org/10.25965/dire.773.

Sousa, R. P., 2017. Agroecologia e educação no campo: Desafios da institucionalização no Brasil, *Educação & Sociedade*, 38, 140, 631–648.

Stassart, P. M., Baret, Ph., Grégoire, J.-Cl., Hance, Th., Mormont, M., Reheul, D., Stilmant, D., Vanloqueren, G., Visser, M., 2012. L'agroécologie: trajectoire et potentiel Pour une transition vers des systèmes alimentaires durables, in Vandam, D., Streith, M., Nizet, J., Stassart, P. M. (Eds.), *Agroéocologie, entre pratiques et sciences sociales*, EDUCAGRI, Dijon, 25–51.

The manufacture of futures and the agroecological transition. Deciphering pathways for sustainability transition in France

Marc Barbier, Sarah Lumbroso,
Jessica Thomas, Sébastien Treyer

With the collaboration of Patrice Cayre,
Roberto Cittadini, Perrine Vandenbroucke
and Jean-Paul Duboeuf

1. Introduction

In the light of various contentions – if not alarms – on the unsustainability of current intensive agricultural systems, intense debates take place in professional, scientific and political arenas on alternative models for agri-food systems (Sumberg et al., 2013). These models are advocated as candidates for transition pathways they could muddle through to achieve a third food regime (Burch et al., 2009). We assist to a proliferation of designations to qualify these new models, such as "ecological modernization", "bio-economy", "sustainable intensification", or "agro-ecology" (Garnett et al., 2013; Levidow et al., 2013; Wezel et al., 2009).

The narratives behind these models carry diverse technical practices but also visions of the future of agri-food systems, described in a more or less comprehensive way. They can then be interpreted as "expectations," meaning "wishful enactments of a desired future" (Borup et al., 2006) or "sociotechnical promises" (Joly, 2013). These activities of building expectations have been classically analyzed as performative, since narratives tend to establish the coordination of actors, the attribution of resources, the delineation of appropriate visions and the engagement of scientific and technological activities (Berkhout, 2006; Chiles, 2013).

Our general assumption is that these expectations entail ontological relationships to change that are also exploration of possible futures grounded in specific areas of experiences and collective expertise. Indeed, they aim at exploring how to change the present to achieve more desirable futures under a pressing state of vulnerability of food system governance (Moragues-Faus et al., 2017). Analyzing the assumptions behind representations of futures can help to decipher how the actors building these representations view the possibilities to combine present and future in actionable transition pathways.

Generally, within debates among stakeholders and in policy debates, controversies between contrasted options are characterized by lack of discussion about the consistent set of assumptions underlying the different transformative pathways that would be requested for sustainability transition of agri-food systems. Moreover, many difficulties arise when the designation of a possible future is supported by and articulated with a critique of the effects of intensive agricultural systems on farmers as well as of the role of economical models in the governance of the "Green revolution" mottos at a distance (Cornilleau, 2016). Today, the diverse representations of the future of agri-food systems do not then spontaneously emerge and co-exist in peaceful debates. Actors allocate resources to the activities of building these narratives on the future of agriculture, and still tensions are high in the political arenas such as in the FAO global dialogue (Loconto and Fouilleux, 2019).

Two archetypes of envisioned transition pathways can be distinguished in the literature about sociotechnical regime in agri-food systems (Barbier and Elzen, 2012; Lamine, 2015; Levidow, 2015): (i) an optimization pathway relying on an improvement in the efficiency of inputs and maintaining an objective of increasing production for food security reasons, which could be aligned with the current organization of the greening of industrial agro-food regime, vs. (ii) a deeper transformation pathway, aiming at reducing the dependence on inputs and leading to more radical and systemic changes based on sustainable criteria and societal values of reconnection. These two archetypal pathways rely on different visions of the relation between ends and means in innovation processes, known as incremental vs. disruptive. The former supposes a rather gradual finalist and deterministic approach of change based on scientific knowledge about what should be sustainable in the end. The latter is more based on an open-ended approach in relation to systemic changes to be more discovered in a state of climatic vulnerability and forthcoming tensions on food commodification and one-health.

In this context, the "deep" agroecological framework seems particularly explicit on the latter as far as both scientific literature and the collective expression of stakeholders to the FAO forum advocate for agroecology as a means to support the realization of key principles (Chappell et al., 2014): attached sustainability to procedural and distributive justice; equitable and active participation of farmers to decision making about food and agriculture; and to foster appropriate local processes of resources management. Agroecology could then be clearly seen as driven by an open-ended perspective liking vision and changes in essential principles of wealth distribution, participation and localism. Nevertheless, this has to be empirically and theoretically challenged (Horlings and Marsden, 2011) since agroecology remains plural and subject to definitional struggles that are also at play in policy making, professional and practitioners facilitation and societal debates.

In this chapter, under the umbrella of recent work in Future Studies claiming for the study of the politics of anticipation (Granjou et al., 2017) and of prospective knowledge practices (McGrail, 2017) for sustainability transition governance, our aim is thus to focus on the links between relationships to change and activities of design of representations of the future regarding the agroecological transition. It implies paying attention to how ontological relationships to change are expressed, mobilized and effective in activities that build representations of the future (e.g. foresights settings, policy instruments and projects) in which agroecology is mobilized as a master frame for transition. The ways in which the representations of the futures are put to discussion can be either grounded in a deterministic approach (providing decision makers with different causality chains between the present and the future among which the issue is to choose the optimal pathway) or an open-ended approach (supporting a collective process of social learning).

2. Analyzing activities of building representations of the future to identify ontological relationships to change

This requires to consider more broadly the works on the links between relationships to change and the design of representations of future that can be found in the literature on the reflexive governance of socio-technical transitions (Smith and Stirling, 2007; Voss et al., 2006), in which specific functions are granted to visions and expectations (Smith et al., 2005). This literature is the prolongation of many attempts to intricate technological forecasts in science and innovation policy (Kemp, 1994; Van der Meulen et al., 2003), in a long march of thinking innovation through national or regional innovation systems. The main rationale there is that the alignment of expectations may be decisive for transition management as they may gather a coalition of actors behind a common objective for change, once a specific momentum for change is achieved in policy making (Hekkert et al., 2007). This rationale can be interpreted as a "projectified" conception of organized action (Sjöblom et al., 2013), assuming that the setting of a policy objective in polities can trigger and organize collective action.

In this chapter we argue that the agroecological transition deserves new ways of analyzing these links between relationships to change and design of representations of futures. It firstly means to view the definition of futures as not only a cognitive process to provide guidance for collective action that take place in foresight or forecast exercise in cycles of innovation in policy making. Activities that define futures can be also directly embedded in collective action, between actors that build a coordinated project. To phrase it differently, the politics of future have agencies and structure at various levels.

In order to frame a broader approach of the links between relationships to change and design activities of futures, we propose the notion of "manufacture of futures" precisely in order to qualify these situated and organized processes of shaping the future in a specific momentum for change. We define the "manufacture of futures" as an organized set of activities that contribute to defining and/or discussing a future representation of a given system, whose perimeters are also a matter of discussion and of definitional struggles in various arenas. It thus encompasses the content as well as the organizational and material aspects of the elaboration of representations of the future that take place in a specific momentum and that target a shift in sociotechnical regime governance. Activities in the manufacture of futures involve the design and prefiguration of transition pathways, based on a more or less explicit ontological relationship to change, meaning the framing of open-ended or deterministic roads to enter the future. Our framework echoes here what Stirling (2008, 2011) had enlightened to define pluralism for the socially balanced assessment of technology.

Through the analysis of the relationships to change built and discussed in the manufacture of futures (Granjou et al., 2017), our objective is to encompass both the processes generally studied in the field of techno-scientific promises that unveils implicit framing and exclusion effects, as well as processes generally studied in the field of futures studies (Slaughter, 2002) that are supposed to aim at structuring a more explicit discussion of all hidden assumptions and open the range of possible futures (Treyer, 2009). The purpose is thus to enlighten the manufacture of future of agroecological transition as they appear in various political arenas. The focus of our enquiry puts then the emphasis on the study of the interplay between the definition of future and relationships to change in terms of governing sustainability transition (Turnheim et al., 2015).

3. The manufacture of futures around the agroecological transition in France

In order to ground this approach of the "manufacture of futures" we use the opportunity of the active debate in France about agroecological transition in the agri-food systems (Bosc and Arrignon, 2020; Compagnone et al., 2018) to analyse jointly and in a common analytical framework, the various activities that produce representations of the future in relation to a specific momentum of sustainability transition labeled "agroecological project for France" in policy-making. The manufacture of futures is distributed in various settings that are related to the agroecological transition policy in France under the current debate on agroecology. Indeed, the agroecological turn is not specific to this country (Ollivier et al., 2019), but agroecology has become there a label of public policy instrument to "farm differently" (Lamine et al., 2020).

Claims for an agroecological transition partly originate from criticisms towards the socio-economic and environmental consequences of the organization of the agro-industrial regime and towards the narrative of neo-liberal productivism supporting this regime (Levidow, 2015). From initial technical issues on the closing of natural cycles and on interactions within biological regulations and thereafter agroecosystems (Altieri, 1983; Gliessman et al., 1981), agroecology has become a wider concept, referring to a discipline, a social movement and a set of practices (Wezel et al., 2009). It questions the organization of the whole agri-food systems to address environmental and social challenges (Gliessman, 2014). All these elements of the debate are present in France.

For some proponents (Duru et al., 2014), an agroecological transition would require a genuine paradigm of change, implying a shift in the relationship between man and nature. These changes would take place in a context of growing tensions between the cultural politics of agriculture models (Meek, 2016) based on epistemic tension about democracy (Carolan, 2008). However, as this concept of agroecology is taken up by a growing number of stakeholders, controversies arise on more or less ambitious operational translations of this concept (Levidow et al., 2014).

Debates on different visions of agroecology are intertwined with debates on the transition pathways that could lead to these visions. In France, the proposal by the Ministry of Agriculture of the "Agroecology Project in France" in 2012 has crystallized debates on more or less ambitious translations of an agroecological transition. This momentum for change is a stimulating and accurate situation for our purpose. It enables to study the production of discourses on the futures of agri-food systems in relation to various situated settings of change for agroecological transition.

4. Methodology and empirical findings

4.1. The portfolio of case studies

Our empirical objective was to characterize how actors involved in the manufacture of futures on the agroecological transition in France conceive their relationships to change of the agri-food systems. We put the emphasis on the analysis of the interplay between the definition of future and relationships to change. Our empirical materials are based on six case studies of different organizational configurations that are related to the political momentum of agroecological transition in France (Tab. 1). They can be divided into three types of activities of future building:

1. Foresight exercises, which are "classical" and explicit activities of building representations of the futures. We analyzed two contrasted projects:

    - The Agrimonde exercise, organized by two French scientific institutions, during which two visions for the future of agriculture at a global scale were built;
    - The Afterres 2050 exercise, organized by an NGO, which ended in the proposal of a land-use change scenario in France;

Tab. 1: Presentation of the portfolio of case studies.

Case study	Type of initiative	Scale	Period	Stakeholders
Agrimonde	Foresight exercise	International	Published in 2009	Experts from INRA / CIRAD (French public agricultural research institutions)
Afterres 2050 Scenario	Foresight exercise	France	First publication in 2009	Solagro Regional authorities, agricultural professionals, environmental NGO, researchers…
The Agroecology Project in France	Political program	France	Since 2013	Ministry of agriculture, local administrative services Agricultural development structures, research and technical institutes, professionals
Ecophyto Action 16	Political program	National	Since 2013	Ministry of agriculture, educational teams in technical high schools
CIVAM group «Empreinte»	Collective project of a group of famers	Local	Since the end of the 1990s	Farmers group CIVAM advisors
TERRAE	Participatory research project for a territorial agroecological transition	Territorial	2013–2018	ISARA-Lyon (research and higher education institute) Local authorities and stakeholders in the Rhône-Alpes region

2. Public policies programs. In those settings, the building of future representations is less explicit, however the policy instruments proposed and the way they are negotiated and implemented are driven by underlying visions of the futures – and relationships to change. Two programs were analyzed:

- the elaboration of a national public policy program, the "Agroecology Project" driven by the French Ministry of Agriculture, which aims at promoting the development of agroecology for the entire French agricultural sector;

- a policy program of future redesign by education, labeled Ecophyto Action 16, which belongs in one of the implementation program of the Agroecology Project. It concerns the technical education system and its aim is to train the future professionals to low-inputs practices. It is notably based on experimentation in technical schools;
3. Local collective projects aiming at a shift of agricultural practices towards agroecology. Two cases fall into this type:
   - a farmers' collective exploring alternative production systems (CIVAM Empreintes);
   - a research-action project (TERRAE) aiming at facilitating an agroecological transition at a territorial scale, involving a diversity of stakeholders.

4.2. Analytical grid

To analyze the relationships to change embedded in these various activities of building the futures, we considered the three following dimensions with the view to decipher the pathways for agroecological transition:

- **The content of the representations of the future that are built.** Analyzing the visions of agroecology and the underlying hypotheses of the transition pathways actually promoted in those representations is a first way to access the relationships to change that are more or less explicitly embedded in the activities of the manufacture of futures.
- **The process implemented to build and discuss these representations of the future.** The methodological tool, the degree of participation of stakeholders (and the types of stakeholders conveyed) can also be analyzed in terms of relationships to frame reflection on changes.
- **The knowledge-ability and activation of practices to enhance the future.** Indeed, the persons in charge of organizing the manufacture of futures do so with specific goals in mind to enhance changes. But the way activity of building representations of the futures and the modalities to discuss them also depend on the type of actionable knowledge produced or expected by collectives.

Those three points are obviously and deeply intertwined in the relationships to change. But we built an analytical grid with questions to address in the analysis of the portfolio of cases studies (Tab. 2).

Tab. 2: An analytical grid of the relationships to change within the manufacture of futures.

Relationships to change through the content of the representations of the futures

What are the relationships to change that are embedded – more or less explicitly – in the representations of the futures built?

– *Materiality of the representations of the future*

What is the degree of formalization of representations of the futures?

Type of future representations: final images, transition pathways?

– *Content of the representations of the future*

Visions of agroecology and degree of transformation of agri-food systems embedded in the representations of the futures

Underlying narratives (and relation with the neoliberal productivist narrative)

Relationships to change through the process of the activity of building representations of the futures

What are the modalities of discussion between the actors involved? How actors influence the process of building and discussing the representation of the futures?

– *Methodological tools for the manufacture of futures*

What tools are used to build and discuss future representations (modelling, participatory workshops…)?

What types of evidence are used to build the credibility of future representations?

– *Organization of actors and spaces for building representations of the futures*

What are the arenas created for the building and discussion of representations of the future?

Which type of actors are involved in these arenas? Are some actors excluded?

Relationships to change through the expected impacts of the knowledge-ability of the futures

What is the type of knowledge produced to enhance changes?

– *Performativity*

How does the organization/team in charge of the activity of building representations of the futures expect it to be performative?

– *Outreach and expected impacts*

What are the objectives for change and how the objectives are expected to be reached?

4.3. Results: Articulations between relationships to change in the activities of the manufacture of futures

Tab. 3 summarizes the main characteristics of the three types of activities of building representations of the futures in our case studies, following the categories of our analytical grid. Regarding the ontological relations to change, an analysis of the content and methodological tools of the case studies makes first reveals a quite clear distinction between deterministic (foresight and public policies) and open-ended perspectives (local projects). However, when looking at the stakeholders and arenas targeted for the building or dissemination of the visions built, and the expected changes of the projects, the perspectives on the relationships to change appear more mixed, especially for the foresight and public policies cases. We first present briefly the three types of activities and then we detail our interpretation of their underlying ontological relations to change in the next section.

Tab. 3: Main characteristics of the activities of building representations of the futures.

Type of activities within the manufacture of futures	Foresight exercises: including the agroecological transition in the range of possibilities for the future of agri-food systems *(Agrimonde, Afterres 2050)*	Public policy elaboration and planning: instituting a pragmatic and inclusive agroecological transition reached by incremental changes *(Agroecology Project, Ecophyto Action 16)*	Local projects: collective and situated explorations for change in practices aiming at deep transformations of the agri-food systems *(CIVAM EMpreintes, TERRAE)*
Content of the future representations manufactured	Formalized future images of agroecological futures (compared with "business-as-usual" images), mainly focused on technical issues but also including other issues (notably food demand and diets)	Public policy discourses, plans and instruments, framing a common and pragmatic horizon for change in agricultural practices. Broad objectives of ecologization of the agricultural sector	Collective actions and experimentations triggered by an ambition for change towards an agroecological transition which is not formalized, with regular redefinitions of objectives and operational actions

Continued

Tab. 3: **Continued**

Type of activities within the manufacture of futures	Foresight exercises: including the agroecological transition in the range of possibilities for the future of agri-food systems (Agrimonde, Afterres 2050)	Public policy elaboration and planning: instituting a pragmatic and inclusive agroecological transition reached by incremental changes (Agroecology Project, Ecophyto Action 16)	Local projects: collective and situated explorations for change in practices aiming at deep transformations of the agri-food systems (CIVAM EMpreintes, TERRAE)
Organization of actors and spaces for the building of representations of the futures	"Ad hoc" arenas of expert knowledge and modelling tools. Discussion with a wider audience and at local scales to improve the images	Public policy building arenas gathering agricultural professionals and some civil society representatives. Asymmetric abilities to influence the decision process among the participants	Collaborative working and experimentation spaces, involving local stakeholders concerned with agri-food systems. An emphasis on the democratic character of discussion processes. It can be farmers led or resting on the initiative of local authorities, civil society organizations, or researchers
The knowledgeability and activation of practices to enhance the future	Building, among recognized experts, robust and credible future images of agroecological agri-food systems. The aim is to disseminate the visions built so that they may thus become a credible option for policy making	Building and adjusting an "acceptable" future so that the maximum of stakeholders engages in projects for change of agricultural practices	Engage local stakeholders in collective experimentation and concrete actions to build "*in itinere*" an agroecological transition of local agri-food systems

5. Interpretation

5.1. The main features of the three types of activities of future building

The first type of activity in the manufacture of futures among our case studies is represented by the two cases of foresight exercises: Agrimonde and Afterres 2050. They are characterized by working arenas composed of experts, created specifically for the manufacture of futures. The aim of these manufactures is to build an argumentation, credible and robust enough to convince and to influence the debates on the future of agri-food systems. By building images of agroecological models fulfilling a set of widely shared objectives, they aim at opening up the range of possible futures, and at making agroecology one desirable and credible option among others. The strength of expertise and techno-scientific evidence to support the demonstration is central in these manufactures.

A second type of activity is characterized by configurations organized by public authorities, which define futures through the framing of agricultural public policies. Those public authorities aim at a wide diffusion of changes in agricultural practices. They make a political proposal of a vision for the future of agri-food systems (e.g. the concept of "double performance", environmental and economical, of agriculture) as an attractor, a political discourse targeting a relatively vague common horizon for change, acceptable by the maximum of actors, with enough implicit assumptions for adjustments and negotiations to remain possible. Then, in the implementation of the political proposal into a policy, the public administrations build a future through regular administrative work, and processes of co-construction of public policies with experts and professionals. The future visions that result are embodied in the forms of public policy instruments, plans and programs, reflecting a "projectified" conception of action. The cases corresponding to this type are the "Agroecology Project in France" (PAE) and the Ecophyto action 16. For this latter case, the implementation of the program rests on experimentations involving teachers and students at the technical school levels.

The third type of activity in the manufacture of futures in our case studies is characterized by participatory processes organized at local levels by collectives with common values, which iteratively define agroecological futures for the concerned territories, mainly based on local experience and knowledge mobilization. This type of manufacture aims at setting the conditions for deep transformations of the agricultural sector and disruptions in current situated practices. The rationale supporting this manufacture is that to provoke potential breakthrough or disruptions, there is a need for concrete exploration of alternative practices. The cases corresponding to this type are the TERRAE research-action project and the CIVAM *Empreintes* initiative.

5.2. Relationships to change behind the visions of the agroecology and the methodological tools used to build them

The two local projects are clearly grounded in a social vision of agroecology, even though the scale envisioned is different: the scale of a farmers' collective for the CIVAM *Empreintes*, the scale of a territory for the TERRAE project. The activity of building visions of the future takes the form of a collective exploration for more sustainable practices, targeting the operational and technical dimensions of professionals' activities, but also socio-economic conditions and organizational principles. Through participatory processes, stakeholders discuss and work on the process of experimentation itself. The priority is not to formalize an explicit future vision. Instead of setting a determined horizon, the flexibility of adaptations nourishing the future vision is supposed to be a guarantee to open up more alternative possibilities. These initiatives thus aim at a participatory elaboration of the transition pathways "*in itinere*". They are therefore clearly grounded in an open-ended relation to change.

The foresight exercises and the public policies programs studied are more grounded in a technical approach of the agroecological transition. The visions of the future built focus on the necessity to overcome techno-scientific challenges, in order to progress toward agroecology, aligned with claims shared by some political, administrative, scientific and professional actors on the need for knowledge production and technical innovation to develop an effective agroecology. The foresight exercises are reflections on the types of technical agricultural systems that will compose the agricultural landscape in the future. Attempts to enlarge the debates to more socio-political issues can be found in the qualitative narrative attached in *Agrimonde* to the quantitative assessments. Indeed, it clearly states that changes in the food system are inevitable in the case of an agroecology scenario. However, the focus put on quantitative evidence in the strategy of the exercise has wiped out this part of the exercise during its dissemination.

In the *"Agroecology Project in France"*, as well as in the *Ecophyto Action 16*, agroecology is associated to a list of technical practices. Precision farming, biocontrol, numeric agriculture or methanization, are examples of some technical practices that compose this framing of agroecology. The methodological tools used in these activities also meet the characteristics of a deterministic approach as described in the introduction of this book: modelling tools based on causality for the foresight exercises; deliberation settings through committees during the elaboration of policy programs to reduce (through negotiation) the diversity of visions for the future of agri-food systems.

5.3. Relationships to change behind the expected impacts of the activities and the stakeholders and arenas targeted to reach them

Another level of analysis of the case studies concerns the outreach and expected impacts of the activities of future building. They are obviously linked with content and method, but we analytically distinguish them as they offer a different perspective on the relationships to change underlying these activities. Indeed, they offer insight on how transition is envisioned in the activities of the manufacture of futures studied.

These conceptions of transition are not clearly described in the content of visions of the future, as no explicit transition pathways are built. For the foresight exercises, only images (at the time horizon of 2050) are proposed. For the public policies programs, objectives are set and technical practices and policy instruments are put forward, without an elicitation of how they can converge in a transition pathway. The local projects are the most explicit ones on this issue of transition, as they imply transition to be built along the process, but the relation to an explicit vision of the future is directly encompassed in action.

The distinction between local projects and the two other types of activities therefore also reflect in two conceptions of how to promote an agroecological transition through the knowledge-ability of futures:

(i) An *"experiential-learning-based transition"*:

Following an open-ended perspective, this conception promotes to start action towards a transition from the experience of local actors, with objectives and methods being jointly and iteratively designed along the way. The arenas of action stay focused around local groups, and don't necessarily intend to reach external decisional arenas or wider debate arenas. They focus on a collective and iterative building of a transition. They can be interpreted as grounded in a regime of "collective experimentation".

(ii) An *"evidence-based transition"*

As for the foresight exercises and the public policy programs, their approach to transition focuses on the role of techno-scientific evidence and argumentation to engage in a transition. The arenas of action are in the field of debates. The main issue is to propose a credible option for change audible in these arenas and able to convince. The conception of transition rests on the assumption that techno-scientific levers and evidence are to be mobilized to convince of the necessity to change and to reach the expected changes. They are closer from a regime of techno-scientific promises. These activities are much more anchored than the previous one in a "projectified" vision of action, in which the setting of objectives (embodied by the formalization of future images supported by techno-scientific evidence in our cases) is supposed to trigger collective action, in a command and control management style. They then refer to a deterministic perspective.

However, the reality is more complex and a closer look at the case studies shows that both types (foresight and policy programs) also have features of an open-ended approach. For the public policies projects, they mix a grounding of argumentation in techno-scientific evidence and in recognition of local and professional experience, through an exemplification of pioneering initiatives. The Ecophyto action 16 case has even more mixed characteristics: its rationale is driven by a techno-scientific argumentation on the necessity to reduce pesticide reduction and by a "projectified" roadmap[1], but its implementation goes by local experimentation in technical high schools in which students and teachers intending to teach transforming practices *in itinere*.

As for the foresight exercises, the relationship to change of their designers can be different depending on the arenas in which they target changes. If the visions of the future that are built (in our cases in approaches that are internal to the organizations launching the projects,

---

[1] As the Ecophyto Plan, in which the Ecophyto Action 16 takes place, aims at reaching a decrease by half of pesticide use in France. In the initial version of the plan, issued in 2009, this objective was set for 2018. This deadline has later been reported.

with no involvement of other stakeholders) are expected to influence decision-makers, by showing one optimal path for the futures among other options, it can be considered a deterministic approach. However, in the cases studied, the arenas targeted are not directly decision arenas, but rather debate arenas. The goal of the projects is to influence debates rather than decisions. Once the visions are produced, the arenas in which they are disseminated gather a diverse audience. Various stakeholders can seize the visions built if they want to use them in their advocacy activities or in local projects. Rather than showing one optimal pathway for the future, the aim is rather to open the field of possibilities, by giving visibility to visions that may be discarded – in our case studies by providing legitimacy and credibility to agroecological practices. This "opening" of the future can be interpreted as being closer to an open-ended perspective on change. Furthermore, those projects can contribute to approaches that are clearly grounded in an open-ended perspective at the local level. For instance, the Afterres2050 scenario has been used in local projects on the transition of food systems.

Conclusion

Our results decipher the ontological relationships to change of actors aspiring to promote an agroecological transition, through the analysis of their activities of "manufacturing of futures". Among our case studies of activities building representations of the future we distinguished three types driven by different rationales: (i) building convincing futures to open the range of options considered in the spaces of debates (foresight exercises), (ii) building a consensual and acceptable future to engage the maximum of actors towards change (public policy programs); (iii) building a future "*in itinere*" through actions grounded in a local context (local participatory projects). The latter is clearly grounded in an open-ended perspective on change, while the others have features of both deterministic and open-ended perspectives.

This articulation between perspectives in these two types of activities, and particularly in the foresight exercises, contributes in our view to enrich reflections on the deterministic vs. open-ended approaches. Indeed, it invites us to go further than the opposition between a « macro, command and control, reductionist » approach and a « local, experimental, participatory » approach, by suggesting there is an intermediary level with perspectives combining criteria of the deterministic approach and the will to open the field of debates, to give space for more deliberative democracy – which tends to be associated with open-ended approaches. Obviously, one should not however overlook the strategies that also exist, in the manufacture of futures, that aim at excluding some visions of the future and lock the debates around one dominant vision for the future (Joly, 2015). Acting at the intermediary level is obviously more than a procedural and diplomatic game, and a too frank opposition between open-ended and deterministic vision is a real *pharmacon*: too much might kill the patient!

Two contrasted – and archetypal – conceptions of the agroecological transition stem from the analysis of the case studies: (i) an "evidence-based" transition and (ii) an "experiential-learning based" transition. These two conceptions could complement themselves: the experiential-learning based transition, particularly because it is grounded in local realities, can be vulnerable to major changes or trends coming from public policy at national or supranational scale (for instance, the common agriculture policy of the EU) or other types of trends in a food system that is always more globalized. With respect to this vulnerability, national or global scale foresights deploy an evidence-based strategy not so much to operate transition, but rather to ensure political and policy space for transition is open for the actors' experiential-learning based transition to happen. Reversely, national or global scale evidence-based transitions often lack a sound base in local action and rather establish the framework conditions for transitions to happen at other scales.

It is striking though that no explicit transition pathways are designed in our case studies. Therefore, the potentially disruptive socio-political choices necessary for an agroecological transition are not raised in the visions of the futures that are built. In these visions, the weakening of intensive farming promoters as a condition for the transition towards more sustainable practices does not appear frankly. The established power relationships characterizing the current agri-food regime are rarely explicitly challenged contrary to what is often expressed in media. Thus the controversial political implications of the agroecological transition remain a blind spot of the manufacture of the futures – at least in our portfolio of case studies. This finding on the incremental and collaborative vision of transition embedded in our case studies leads us to suggest that the framework of deterministic vs. open-ended approach could be complimented by an analysis of collaborative vs. adversative relationships to change, exploring the complementarity of consensual or dissensual visions in a pluralistic approach of the manufacturer of futures. Such a matrix could then reflect how agroecological transitions dialectically deal with the sub-politics of framing robust discontinuation in the conventional regime.

References

Altieri, M. A., 1983. *Agroecology: The Scientific Basis of Alternative Agriculture*. Published by the Division of Biological Control, University of California, Berkeley.

Barbier, M., Boelie E., 2012. *System Innovations, Knowledge Regimes and Design Practices towards Transitions for Sustainable Agriculture*, Paris, INRA Edition.

Berkhout, F., 2006. Normative Expectations in Systems Innovation, *Technology Analysis & Strategic Management*, 18, 3–4, 299–311.

Borup, M., Brown, N., Konrad, K., Harro Van Lente, H., 2006. The Sociology of Expectations in Science and Technology, *Technology Analysis & Strategic Management*, 18, 3–4, 285–298.

Bosc, C., Arrignon, M., 2020. *La transition agro-écologique en France ou les conditions du changement agricole*, Rennes, Presses Universitaires de Rennes.

Burch, D., Lawrence, G., 2009. Towards a Third Food Regime: Behind the Transformation, *Agriculture and Human Values*, 26, 267–279.

Carolan, M. S., 2008. Sustainable and Conventional Agricultural Field Days as Divergent Democratic Forms, *Science, Technology, & Human Values*, 33, 4, 508–528.

Chappell, J., et al., 2014. Scientists' Support Letter for the International Symposium on Agroecology, 18–19 sept. 2014. [online] URL: https://www.iatp.org/.

Chiles, R. M., 2013. If They Come, We Will Build It: In Vitro Meat and the Discursive Struggle over Future Agrofood Expectations, *Agriculture and Human Values*, 30, 4, 511–523.

Compagnone, C., Lamine, C., Dupré, L., 2018. La production et la circulation des connaissances en agriculture interrogées par l'agro-écologie. De l'ancien et du nouveau, *Revue d'anthropologie des connaissances*, 12, 2, 12–24.

Cornilleau, L., 2016. La modélisation économique mondiale, une technologie de gouvernement à distance? *Revue d'anthropologie des connaissances*, 10, 2, 171–196.

Duru, M., Fares, M., Therond, O., 2014. Un Cadre Conceptuel Pour Penser Maintenant et Organiser Demain La Transition Agroécologique de L'agriculture Dans Les Territoires, *Cahiers Agricultures*, 23, 2, 84–95.

Garnett, T., Appleby, M. C., Balmford, A., et al., 2013. Sustainable Intensification in Agriculture: Premises and Policies, *Science*, 341, 6141, 33–34.

Gliessman, S. R., 2014. *Agroecology: The Ecology of Sustainable Food Systems*, Third Edition. CRC Press.

Gliessman, S. R., Garcia, R. E., Amador, M. A., 1981. The Ecological Basis for the Application of Traditional Agricultural Technology in the Management of Tropical Agro-Ecosystems, *Agro-Ecosystems*, 7, 3, 173–185.

Granjou, C., Walker, J., Salazar, J., 2017. The Politics of Anticipation: On Knowing and Governing Environmental Futures, *Futures*, 92, 5–11.

Hekkert, M. P., Suurs, R. A. A., Negro, S. O., Kuhlmann, S., 2007. Functions of Innovation Systems: A New Approach for Analysing Technological Change, *Technological Forecasting & Social Change*, 74, 413–432.

Horlings, L. G., Marsden, T. K., 2011. Towards the Real Green Revolution? Exploring the Conceptual Dimensions of a New Ecological Modernisation of Agriculture that Could 'Feed the World', *Global Environmental Change*, 21, 2, 441–452.

Joly, P. B., 2013. On the Economics of Techno-Scientific Promises, in Akrich, M., Barthe, Y., Muniesa, F., Mustar, P. (Eds.), *Débordements : Mélanges Offerts a Michel Callon*, Sciences Sociales. Paris, Presses des Mines, 203–221.

Kemp, R., 1994. Technology and the Transition to Environmental Sustainability –The Problem of Technological Regime Shifts, *Futures*, 26, 10, 1023–1046.

Lamine, C., 2015. Sustainability and Resilience in Agri-Food Systems: Reconnecting Agriculture, Food and the Environment, *Sociologia Ruralis*, 55, 1, 41–61.

Lamine, C., Derbez, F., Barbier, M., 2020. L'indétermination performative d'instruments d'action publique pour la transition agroécologique, in Bosc, C., Arrignon, M. (Eds.), *La transition agro-écologique en France ou les conditions du changement agricole*, Rennes, Presses Universitaires de Rennes.

Levidow, L., 2015. European Transitions towards a Corporate-Environmental Food Regime: Agroecological Incorporation or Contestation? *Journal of Rural Studies*, 40, 76–89.

Levidow, L., Birch, K., Papaioannou, T., 2013. Divergent Paradigms of European Agro-Food Innovation the Knowledge-Based Bio-Economy (KBBE) as an R&D Agenda. *Science, Technology & Human Values*, 38, 1, 94–125.

Levidow, L., Pimbert, M., Vanloqueren, G., 2014. Agroecological Research: Conforming – or Transforming the Dominant Agro-Food Regime? *Agroecology and Sustainable Food Systems*, 38, 10, 1127–1155.

Loconto, A. M., Fouilleux, E., 2019. Defining Agroecology, *The International Journal of Sociology of Agriculture and Food*, 25, 2, 116–137.

McGrail, S. D., 2017. *The Roles and Use of Prospective Knowledge Practices in Sustainability-Related Transitions: A Realist Evaluation and Pragmatist Synthesis*, Sydney, (Doctoral dissertation). [online] URL: https://opus.lib.uts.edu.au/bitstream/10453/123188/2/02whole.pdf.

Meek, D., 2016. The Cultural Politics of the Agroecological Transition, *Agric Hum Values*, 33, 275–290.

Moragues-Faus, A., Sonnino, R., Marsden, T., 2017. Exploring European Food System Vulnerabilities: Towards Integrated Food Security Governance, *Environmental Science & Policy*, 75, 184–215.

Ollivier, G., Bellon, S., de Abreu Sá, T. D., Magda, D., 2019. Aux frontières de l'agroécologie. Les politiques de recherche de deux instituts agronomiques publics français et brésilien, *Natures Sciences Sociétés*, 27, 1, 20–38.

Sjöblom, S., Löfgren, K., Godenhjelm, S., 2013. Projectified Politics – Temporary Organisations in a Public Context, *Scandinavian Journal of Public Administration*, 17, 2, 3–12.

Slaughter, R. A. (Ed.), 2002. *New Thinking for a New Millennium: The Knowledge Base of Futures Studies*. London, Routledge, First Edition, 260 p.

Smith, A., Stirling, A., 2007. Moving Outside or Inside? Objectification and Reflexivity in the Governance of Socio-Technical Systems, *Journal of Environmental Policy & Planning*, 9, 3–4, 351–373.

Smith, A., Stirling, A., Berkhout, F, 2005. The Governance of Sustainable Socio-Technical Transitions, *Research Policy*, 34, 10, 1491–1510.

Stirling, A., 2008. « Opening Up » and « Closing Down », Power, Participation and Pluralism in the Social Appraisal of Technology, *Science Technology Human Values*, 33, 2, 262–294.

Stirling, A., 2011. Pluralising Progress: From Integrative Transitions to Transformative Diversity, *Environmental Innovation and Societal Transitions*, 1, 1, 82–88.

Sumberg, J., Thompson, J., Woodhouse, P., 2013. Why Agronomy in the Developing World has Become Contentious, *Agriculture and Human Values*, 30, 71–83.

Treyer, S., 2009. Changing Perspectives on Foresight and Strategy: From Foresight Project Management to the Management of Change in Collective Strategic Elaboration Processes. *Technology Analysis & Strategic Management*, 21, 3, 353–362.

Turnheim, B., Berkhout, F., Geels, F., Hof, A., McMeekin, A., Nykvist, B., van Vuuren, D., 2015. Evaluating Sustainability Transitions Pathways: Bridging Analytical Approaches to Address Governance Challenges, *Global Environmental Change*, 35, 239–253.

Van der Meulen, D. J., Rutten H., 2003. Developing Futures for Agriculture in the Netherlands: A Systematic Exploration of the Strategic Value of Foresight, *Journal of Forecasting*, 22, 219–233.

Voss, J.-P., Bauknecht, D., Kemp, R. (Eds.), 2006. *Reflexive Governance for Sustainable Development*, Cheltenham, UK, Edward Elgar.

Wezel, A., Bellon, S., Doré, T., Francis, C., Vallod, D., David C., 2009. Agroecology as a Science, a Movement and a Practice. A Review, *Agronomy for Sustainable Development*, 29, 4, 503–515.

How access and dynamics in the use of territorial resources shape agroecological transitions in crop-livestock systems: Learnings and perspectives

Vincent Thénard, Gilles Martel,
Jean-Philippe Choisis, Timothée Petit,
Sébastien Couvreur, Olivia Fontaine, Marc
Moraine

1. How territorial resources and local dynamics support agroecological transitions

1.1 Resources involved in agroecological transitions

The multi-level perspective (Geels, 2004) is widely used to define sociotechnical transitions through the sociotechnical landscape's pressures. This perspective raises existing problems in the dominant regime, and innovations in niches (Geels and Kemps, 2007). Agriculture here is conceptualized as a patchwork of sociotechnical systems embedded in various trajectories of evolution. Among them, agroecological transitions correspond to multiple processes starting from niches supporting "radical, systemic changes" in social, technological, political or institutional areas, but it is also the result of collective action for building and sharing knowledge (Elzen et al., 2017). These changes are based on individual and collective strategies supported by various resources: access to land, infrastructures, and institutional or informal networks (Wezel et al., 2009).

According to Buclet and Cerceau (2019), a territories' sustainable development depends on optimization and distribution in the use of material, immaterial and financial resources. We consider four types of resources:

(i) Natural resources supporting ecological processes in production systems: land, water, animal and vegetal biodiversity;
(ii) Technical and cognitive resources influencing ecological processes through adequate practices: specific equipment, adapted breed or crop varieties, specific know-how, farmers' ability to manage complexity, and trade-offs between short *vs.* long term benefits;
(iii) Social resources sustaining agroecological systems' legitimacy and recognition: social networks, local support from diverse stakeholders;
(iv) Economic resources enhancing the viability in agroecological systems: marketing channels, public subsidies.

Local, territorial and even global contexts can provide these resources. Natural ones are related to local conditions, while technical ones can be generic but must be locally adapted. Social and economic resources can be structured on national or supranational scale (NGOs, trade rules, policies), but depend on local networks and implementations. The role of local stakeholders and coordinated actions appears essential in the activation of territorial resources (Colletis and Pecqueur, 2005). Along with agroecological transitions, resources support the establishment and resilience of agroecological livestock farming systems (LFS) in the adaptation to hazards (Milestadt et al., 2012). Madelrieux et al. (2017a, b) identify the possible synergies between production systems, for example, when LFS use local biomass and semi-natural spaces, and provide effluent used by other activities. According to Rigolot et al. (2019), LFS' adaptation ability is a key element for the struggle with climatic and sanitary risks affecting the growth and quality of feed resources; market fluctuations in products and inputs; institutional risks related to regulation or policy changes; financial risks; and human-related risks (diseases, accidents, disabilities). We describe varied combinations of resources supporting LFS, and stakeholders' influence in resources mobilization.

1.2 How agroecological LFS combine biodiversity and territory embeddedness

Let us consider LFS's agroecological features according to Therond's two-dimensional approach (Therond et al., 2017): on one hand the mobilization of biodiversity in the productive process; and on the other hand LFS's territorial embeddedness regarding supply and commercialization chains. We identify dynamics of change within LFS moving towards agroecology, considering resources combination in different LFS.

Biodiversity mobilization's intensity can be estimated according to four criteria:

- Diversity of land uses;
- Diversity of reared species;
- Type of animal bred (breed and mode of selection);
- Management contribution to natural areas or specific ecosystems.

The intensity of local food chain anchorage is also measured according to four criteria:

- Activities' diversification: the nature of farms' activities, pluriactivity occurrence;
- Local production processing and marketing: farm shops or local distribution chains;
- Local purchase of inputs: inputs' origins and nature, purchase frequency;
- Collective dynamics on a local level, governance and shared values: stakeholders from the same local networks co/managing their own governance structures; according to share representations of common values and objectives in relation with land.

2. Looking at agroecological transitions in contrasting French territories

Let us consider transition in LFS through four contrasted cases from four different French areas. The differences of climate, ecosystems, and socioeconomic dynamics, determines some of territorial resources' various aspects. Two archetypal LFS define each of the four different territories: a "baseline", often a conventional system broadly spread, and an "agroecological niche" which represents an advanced and promising archetype of agroecological transition.

2.1 Territories description regarding territorial embeddedness and biodiversity mobilization

Mediterranean area in Languedoc

Languedoc is a patchwork of plains, scrublands and middle height mountains. It is a dry and windy area with Mediterranean climatic conditions. Rainfalls (600–1200 mm) are concentrated on heavy rain episodes and mostly shallow, rocky and poor soils.

Wine industry is located in the lowlands while specialized livestock farms are in mountain areas. Pastures, arable lands and vineyards represent the Usable Agricultural Area (UAA). On average in the region, farms are rather small (29 hectares of UAA), mainly due to small vineyards, fruit and vegetable farms. The main breeding system is agropastoralism, based on extensive grazing and feed inputs; and consequently these farms are larger. This territory has to deal with a dangerous exposure to climate change. Indeed, recent years' frequent and severe droughts have affected fodder resources availability threatening LFS's sustainability. We focus on mixed LFS as different in context as plains ones and hills-based ones, with various combinations of species including ruminants and monogastrics (Fuselier, 2019).

The baseline "Pig-Goat system – PGS" (UAA 70–400 ha, Organic farming and/or products with PDO label) relies on a network of mixed livestock farms combining goats for cheese production and pigs raised outside with by-products of cheese processing, and diversified crops.

The agroecological niche "Mixed Rangeland system – MRS" (UAA 90–1000 ha, Organic farming; robust breeds) defined by mixed livestock farms combines various herbivores and monogastrics. This niche only uses local resources and performs a high level of self-sufficiency. Both LFS types are engaged in agroecological transition, but the niche is above the baseline.

Territorial embeddedness is strongly guaranteed by a short supply chain and local economic empowerment. Diversified products are sold through direct sales or short supply chains (farm or local shops) which improves production added value and forges the bond with local consumers/citizens. Some farmers sell part of their products through long supply chains but always with a PDO label or niche markets such as luxury groceries or restaurants. Farmers often organize local supply for their feed inputs (hay from the area, or further like *Crau* hay about 200 km away). MRS use mainly local or regional by-products (*Camargue*'s rice straw, unsold fruits and vegetables from local shops). Whereas PGS can buy more standard feed (rapeseed cake) in remote areas, especially to ensure milk production for dairy ewes or goats. Pigs play an important role because they use local feed resources: whey from cheese production, acorn from wooded areas. Many of the farmers develop complementary activities, like hosting tourists or school classes, and provide their territory with services such as contracts for grazing against forest fires.

Biodiversity mobilization is higher for MRS than for PGS. MRS utilizes grazing on semi-natural pastures and rangelands of high nature value (Natural Parks, Natura 2000). The multiple livestock species allow to manage grazing in heterogeneous rangelands. Farmers often contract with biodiversity protection actors (Conservatories for natural areas, BirdLife NGO, etc.) to preserve patrimonial species (e.g. *Griffon vulture*, peat bogs' *Drosera*) or control invasive species (e.g. seagrass, *Russian olive tree*). Such cooperation can be rewarded directly or can be a commercial advantage. The choice of animal breed and the management of reproduction is carefully adapted to local constraints. While traditional sheep breeds (*Caussenarde des Garrigues, Raïole*) are still raised, *Highland* or *Galloway* cattle have been imported from United Kingdom for their rustic features well adapted to the harsh conditions of grazing in wetlands. In PGS, grazing areas and fodder sources are more frequent (permanent or temporary grasslands). Animal selection is often a combination of common and hardly breeds (e.g. crossbreeding *Duroc* pig with *Porc Noir Gascon*).

Oceanic area in Brittany

Brittany's livestock density is high. Climatic conditions and soil fertility, combined with land flatness guarantee a good growth of grass but also good yields on crops. Since the 1960s, the city of Rennes is the most densely populated area specialized in dairy production. Several international firms are implanted and trade on international markets. Strong supply chains and advisory services support high productivity livestock development with confined animals fed with maize and purchased concentrates. Environmental issues such as nitrate pollution and land degradation led to the development of alternative systems based on grasslands. Meanwhile the proximity of the city offers opportunities to develop organic production and local food network. Local policies aim to keep agriculture on the area and set up ecological networks (Couvreur et al., 2019; Petit et al., 2019a, b).

The baseline is "Maize dairy cow system – MDS" (UAA 40–180 ha, 7000–10500 kg milk per cow, mainly *Holstein* breed). Farmers look for higher work productivity, they develop cereal production and aim at increasing milk quantity per cow and stocking rate (above 1.7 LU.ha^{-1}).

The agroecological niche is "Grassland dairy cow system – GDS" (UAA 50–110ha, 3500–8500 kg milk per cow, mainly Holstein breed). Producers look for feed self-sufficient farms based on grass management (1.4 LU.ha^{-1}).

Territorial embeddedness is higher with GDS than MDS. The latter being the legacy of decades of structuring industrial dairy sector: genetic selection (*Holstein* breed), animal feed companies, product packaging, slaughterhouses, agronomic research (INRA experimental station located 8 km from Rennes), advisory structures and technical institutes. Maize and soybean meal have been introduced since the 1960s, along with *Holstein* breed, requiring high feed inputs. Feed production companies value cereals produced by dairy farmers and integrate imported soybean to produce cheap feed. MDS products are processed in the industrial sector, and exported outside Brittany (the rest of France and world markets). Like elsewhere, farm size increases and new areas are mainly devoted to increasing crop production.

GDS grew out of the Sustainable Agriculture Network in the mid-1990s. Alternative Farmers' Federation has given rise to grass-based farms that limit inputs and aim to achieve feed self-sufficiency and/or organic farming. A part of the advisory sector specializes in grassland seed (especially diversified grass and legume associations) and cover crop management methods. The inhabitants of Rennes are a good clientele for products from these farms, either through direct sales or by purchasing local organic products.

Biodiversity mobilization is medium in GDS and low in MDS. This latter ensures forage production with chemical inputs providing high milk yield, and concentrates purchase compensates the energy/protein imbalance. Environment is seen as a factor to be controlled. Nevertheless, crop rotation is diversified with winter and spring crops and temporary grasslands (mainly ryegrass and white clover). Crops are organized in 5-year rotations with pasture around the farm head, while distant plots can be dedicated to crop production as wheat, barley, rapeseed.

Holstein breed is the most common, embryo sexing occurs regularly, the main selection criterion being milk quantity of milk.

GDS also raises *Holstein* but here, selection criteria are multiple (longevity, milk quality, fertility, etc.), even sometimes carrying out crossbreeding. Meadows are the main cover and farmers adapt the sowing to soil characteristics. There are grasslands with fescue, others with orchard grass and also multi-species grasslands commonly grown in association with legumes, including clovers. Preserving grasslands as long as possible and developing patches of permanent meadow providing a heterogeneous landscape is a constant concern.

Semi-continental area in Aveyron

Aveyron is a southern Central Mountain region. The most original animal production being sheep milk, traditionally meant for "Roquefort" cheese production, especially famous for having been the first to obtain Protected Denomination of Origin (PDO) certification in 1925. For a long time milk production was limited to low-productivity limestone uplands called *Causses*. Traditional dairy sheep breeding has been modernized since the 1970s. Today's sustainability challenge is feed self-sufficiency improvement, targeted through intensification of meadows, native grassland use, milk production cuts, and milk price boost (Thénard et al., 2014; 2016, 2018).

The baseline "Foddering dairy sheep system – FSS" (UAA 35–235 ha, 215–375 kg milk per ewe) has a high level of animal productivity. Usually, milk is produced in winter and spring. Plant resources diversity is wide and farmers cultivate sown pastures for grazing and harvesting. This baseline with smaller farms and higher animal productivity is located in the western zone. Farmers frequently harvest grass silage.

The agroecological niche "Grazing dairy sheep system – GSS" (UAA 65–538 ha, 170–300 kg milk per ewe) is located in the southern zone with harsher agronomic conditions: drought in summer and a light soil. Farmers use more grazing and match milking period with grass growth. In summer, farmers use rangelands or wooded pastures to limit forage consumption. Many farmers are leaving the "Roquefort" PDO organization to free themselves from production constraints, developing alternative production systems and commercialization chains.

Territorial embeddedness is medium due to local "Roquefort" cheese production, but now partly standardized and industrialized. The notion of *terroir* determines PDO label and mobilizes both local biological resources (*Lacaune* sheep breed), natural caves for the maturing of cheeses and human know-how. Today, a high level of rural employment still depends on sheep's milk production. FSS has spread to the rich lands of the *Segala* in western *Aveyron* with high productivity based on feeding with grass silage, soybean meal and dehydrated alfalfa produced in *Champagne* region. Multinational companies collect the main part of milk production. Some of it is used for "Roquefort" cheese (45%), with most of it is processed into industrial products without quality signs, sold in globalized supply chains. Farmers' organizations have initiated a process of reflective thinking on technical and commercial alternatives to deal with this situation. GSS on the *Causses* produces milk mainly in summer and autumn, which is processed into yoghurts or local cheeses through a local cooperative. Agricultural development is very active in this region; livestock farmers are mobilized in the challenges of enhancing their territory's value, searching for local resources autonomy and added value.

Biodiversity mobilization is variable. Animal genetic resources are limited as only the *Lacaune* breed is authorized in accordance with "Roquefort" specification. The challenge is to renew the selection criteria to better adapt to the agroecological transition and increase genetic diversity. Plant resources diversity is lower in FSS than in GSS. All arable land had been ploughed for many years and forage crops are abundant. Intensification using nitrogen fertilizer has led to short-term monospecific and intensive sown meadows (Italian and hybrid ryegrass, red clover). Forage crops such as alfalfa are also very common, sometimes combined with grasses. Forage production objectives are mainly a higher degree of intensification for FSS and longer-term grasslands for GSS. Recently, repeated summer droughts have impacted grassland sustainability especially in GSS, pushing farmers to increase diversity of meadows and crops (multi-species mixture, selection of local seeds, cereal-legumes associations). In various areas where tillage is impossible (rugged areas, wetlands near the rivers, calcareous rangelands) GSS maintains natural grasslands, rich in biodiversity. Sloped woodlands are also sources of grazing, especially in summer or winter. Some of these natural resources are included in the UNESCO World Heritage area and as Natura 2000 areas for the wealth of their fauna and flora. The resurgence of wolves on the *Causses* can be a source of concern.

Tropical area in Reunion Island

In *Reunion Island*, animal production serves to employ people and to contribute to food autonomy. Animal farming has developed in organized cooperative sectors that cover 24% (beef meat) to 94% (eggs) of local consumption. An important part of subsidies is dedicated to help farmers, cooperative and agro-industries increase productivity and compensate huge production costs due to the island remoteness. Because of agricultural area scarcity, LFS have been set up according to intensive models and many inputs. These models are hardly compatible with organic farming and the elevated price of animal products is not an incentive to change. However, cow, calf and goat breeding are important users of local forage resources (grazing and mowing). Goat production is mostly a complementary activity, representing 10–50% of the total income. There are different systems, from multi-active or backyard farmers with few animals to large herds (30–100 goats) (Fontaine et al., 2010). Most of professional farms mix crop culture (sugarcane, market gardening, and arboriculture) with animal breeding (bovine meat, pig, poultry).

The baseline "Tropical Mixed livestock system – TMS" (UAA 5–60 ha, 20–100 goats-*Boer* breed), is based on a mixed indoor x grazing system. This is conventional farming with diversified production: meat (goat, cattle, pig, poultry) and crops (sugarcane, meadows, etc.).

The agroecological niche "Tropical Garden livestock system – TGS" (UAA 1–8 ha, 10–30 goats *Creole x Boer* breed) is based on small farms gardening and rearing goats for income diversification and manure availability. TGS intents improving its self-sufficiency.

Territorial embeddedness: Goat breeding is a diffuse animal husbandry rooted in the local society. Goat animals or meat (including cull goats) is a highly sought-after production sold through local food networks (direct sale and traditional butcheries). It is the basis of the island's traditional dish (*cabri massalé*), and a product for ceremonial slaughter (e.g. for the Tamil community needing well-conformed goats) (Fontaine et al., 2008).

Favorable valorization of animals allows feed purchase (hay and concentrate). There is a great diversity of practices and systems, from self-sufficient with low use of concentrate and veterinary products, to high dependency. Breeders having small agricultural areas use natural forage resources from outside of the farm (mowing grass on roadside, grazing wasteland and savannah) or buy hay (to lowlands' producers). Some systems are, however, highly dependent of imported concentrates (especially for fattening male goats). Breeding sector is mainly organized in the cooperative model with slaughtering and manufacturing facilities to ensure mass retailing. In this system, goat meat is not profitable enough because imported frozen meat is cheaper; recently a goat cooperative was forced to close. Farmers are weakly organized for marketing products with high demand.

Biodiversity mobilization: In comparison to other productions developing on a "technological package" pattern, goats are using the greatest diversity of forage resources with (i) natural resources, (ii) cultivated resources mowed (elephant grass, temporary meadows for hay) or grazed (pasture), (iii) forage crops (maize) and (iv) by-products (sugarcane tops and straw). Farms with several species of livestock are common as sheep and goats are a diversification form for the rearing of cattle or monogastrics (pigs or poultry). The local goat breed *Pei* is hardy (good mothering abilities, prolificacy, etc.), and is well suited to the environment. However, this breed with small size does not correspond to current expectations (well-conformed goats for ceremonial slaughter). *Boer* breed was imported from South Africa and produce better conformed goat. Finally, goat population is mainly crossbreeding *Pei x Boer*.

Goats are mostly bred in housing, with trough fed with green fodder or hay supply. The small size of the herds, diversification and multi-activity complicate grazing practices all the more because of the constant monitoring required by stray dogs or animal thefts issue. However, grazing is applied for large herds, on cultivated meadows or on natural resources. A research project is underway to find an opportunity to manage savannah with cattle and goat grazing.

2.2 Agroecological LFS multi-criteria assessment

Assessing LFS agroecological performances requires surveying complex ecosystem functions. For this, normative methods such as Life Cycle Assessment (de Vries and Boer, 2010) are less relevant than multi-criteria approaches studying jointly natural resources, ecosystem services, health management, socioecological resilience, etc. (Affholder et al., 2019). We focus on the agroecological functioning dimensions that are directly connected to farmers' practices (Thénard et al., 2016; Magne et al., 2019). We thus assess LFS through four dimensions and nineteen corresponding indicators (Fig. 1). This set of indicators determines the extent to which LFS implement the principles of agroecology by mobilizing biodiversity-based processes, recycling energy and nutrients, improving diversity and connectivity on the farm and on local territory (Dumont et al., 2013; Bonaudo et al., 2014; Thénard et al., 2014). Best scored are LFS which boost agroecological management principles: soil fertility increase, chemical inputs restriction, farm autonomy improvement, integrated crop management and animal diversity promotion.

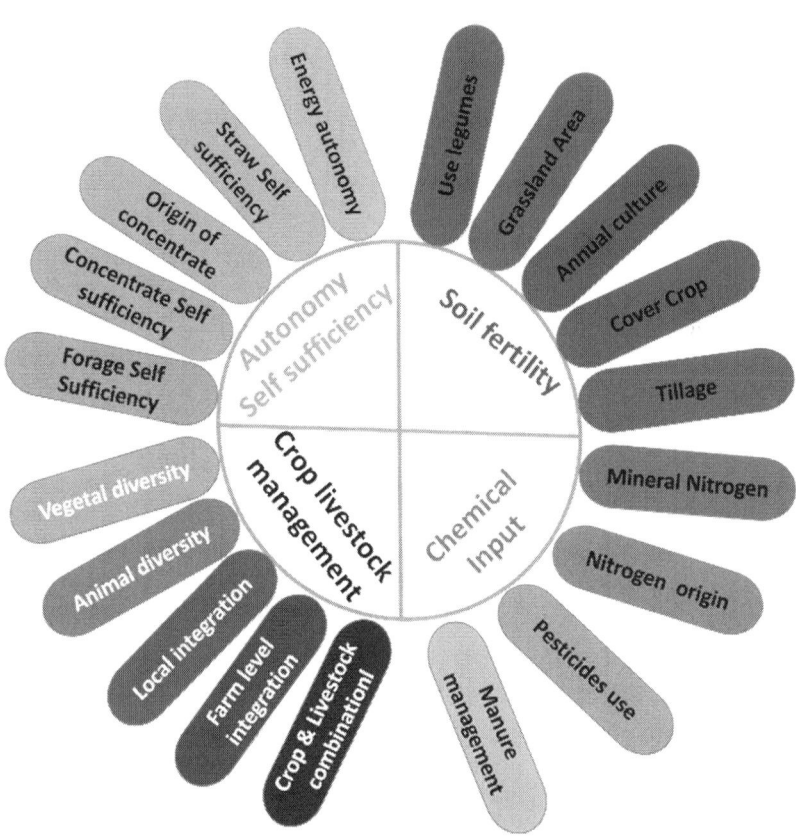

Fig. 1: Indicators used to evaluate the LFS according to the four agroecological performances. Adapted from Magne et al. 2019.

3. From mobilized resources to LFS agroecological performances

3.1 LFS archetypes in case studies

Inspired by Therond et al. (2017), we have placed the eight LFS archetypes on a factorial map (Fig. 2) according to the four types of resources (cf. §1.1) mobilized by each one. The horizontal axis represents the system's territorial embeddedness and the vertical axis shows biodiversity integration level. Archetypes draw a diagonal from less (MDS) to more (MRS) anchorage in the territory and biodiversity-based. Despite differences of resources used between regions and systems, this representation allows positioning the production systems in an agroecological gradient. For each territory, niche systems – all located in the right and top quarter- logically turn out more advanced in terms of biodiversity integration and territorial embeddedness than the baseline, but significant differences

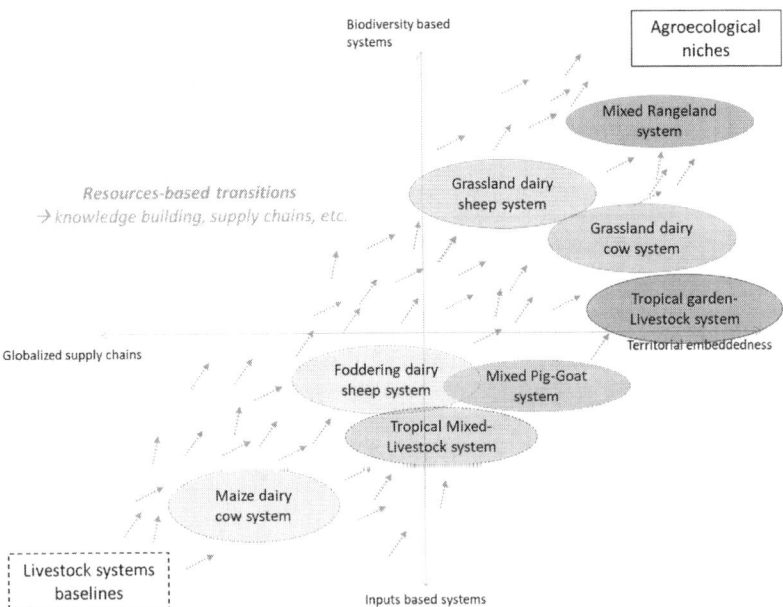

Fig. 2: Position of the eight LFS archetypes on a factorial map. The dotted arrows show possible pathways for transitions towards agroecological systems, independently from any unique roadmap. Adapted from Therond et al. 2017.

also appear within and between regions, underlying that trajectories and targets are context dependant. In order to understand how the systems have been placed on the map, we will study resources used in each case study and by each archetype in the next paragraph.

3.2 Mobilized resources in the case studies

LFS always rely on a combination of natural (ecosystems), technical (animal breed and management skills), economic (markets) and social (networks, support from local stakeholders) resources (Figs. 3, 4, 5, and 6).

Languedoc's mixed farming systems combine nature-oriented ecosystems, strong products valorisation and knowledge-sharing networks.

In *Languedoc*, both types of LFS use of rangelands, scrublands and more (MRS) or less (PGS) natural grasslands (Fig. 3). They rely on different social networks to access grazing areas, technical advice or local support: farmers' associations, local stakeholders such as forest

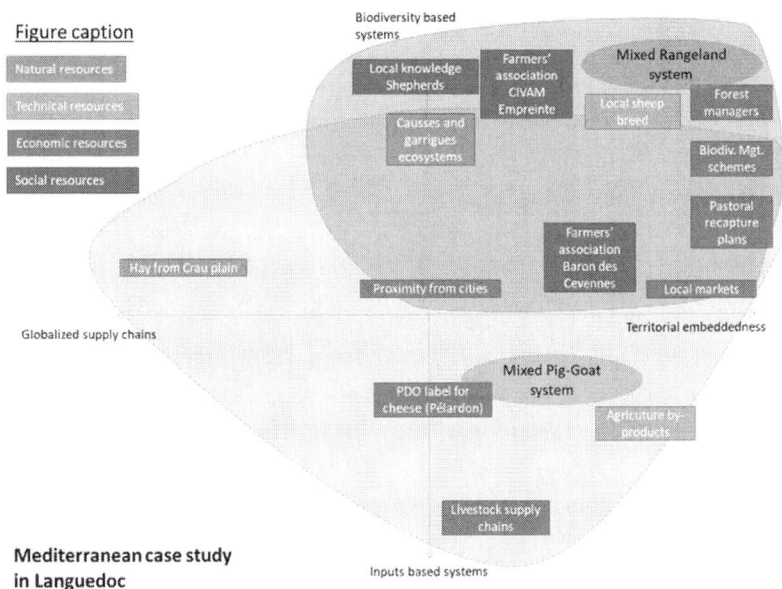

Fig. 3: Resources mobilized in the Languedoc case study.

managers or nature conservation societies. Farmers partly or fully sell their products through direct sales or local food networks, thus increasing benefits (Fuselier, 2019). Besides, subsidies from the European Common Agricultural Policy (CAP) largely contribute to their income, together with additional supports from territory stakeholders including indirect economic gain such as free access to grazing areas, or rewarded environmental services. Local stakeholders legitimize their LFS communicating on their importance to manage ecosystems and to preserve biodiversity.

The social networks linked with these two LFS permits the exchange of experience and knowledge, and increase collective support and share objectives.

The *Baron des Cévennes* society helps marketing pig products (ham and sausages) of PGS which is more oriented on commercial purposes. The center of rural initiatives (CIVAM) *"Empreinte"* promotes the defense of common values and a specific vision of the profession. Both groups also exchange resources such as breeding animals, equipment, work for organizing transhumance, etc. PGS sells *"Pélardon"* PDO cheese in long supply chain and need to use resources less embedded in local territory, like hay from the *Crau* plain (150 km away), feed from suppliers, but also a part of external inputs, like by-products, supplied by territory (e.g. unsold fruits or vegetables fed to the pigs).

Brittany's dairy systems promote intensive LFS products to local consumers with labels.

In *Brittany*, oceanic climate and good soil fertility are favorable to all kinds of resources for both systems (Fig. 4). On one hand, MDS produces maize needing water, on the other hand GDS promotes grass growth in spring and autumn. However, MDS has also a high use of feed from local companies with imported concentrate and by-products of the local agro-food sector. Both systems integrate genetic *Holstein* breed selection, but with different selection criteria. MDS targets milk high quantity and fertility whereas GDS looks for milk quality, robustness of cows and fertility. Both systems can use the *"Bleu Blanc Coeur"* label as an economic resource based on milk composition without specification of practices. For that, MDS uses inputs such as extruded flaxseed, while GDS achieves milk composition thanks to grass use.

Resources for agroecological transitions in livestock

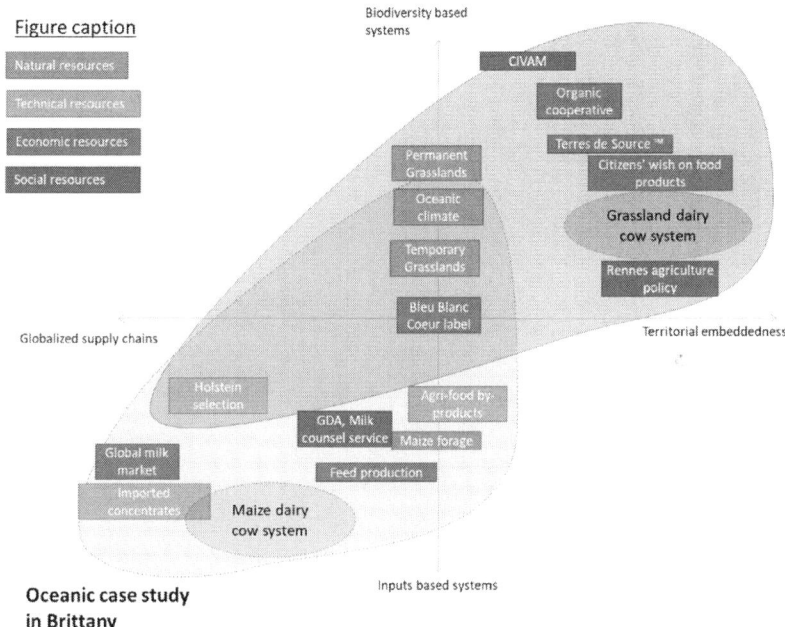

Fig. 4: Resources mobilized in the Brittany case study.

MDS milk is mainly produced for the world market (milk powder, cheese, conditioned milk). GDS milk can be sold through cooperatives specialized in organic agriculture or through local food network thanks to Rennes' inhabitants demand. The city recently launched a new specification called "*Terres de sources*." The brand is based on a sustainability score which must be greater than a minimum threshold, and the commitment to improve its value over a period of 5 years. This new possibility of promotion can be seized by both systems but the expected score is better achieved by GDS than MDS.

Finally, the two systems mobilize different consulting networks. CIVAMs support GDS' farmers by providing advice on grass management and low-input dairy production, while mainstream advisory bodies provide advice on milk productivity, fertilizer use or legumes integration to complement maize silage.

Aveyron's dairy sheep systems combine intensification of milk production and feed self-sufficiency thanks to a wide variety of plant resources and a strong network for farmers' advice.

In *Aveyron*, vegetal diversity is important and "*Roquefort*" PDO label encourages local resources use (Fig. 5). The intensification of milk production has led to an increase in inputs purchased mainly on the world market (nitrogen fertilizers, seeds, soybean meal, and dehydrated alfalfa) which are used by both systems. However, GSS farmers are trying to improve feed self-sufficiency and therefore use more local resources produced from diversified grasslands and native grasslands.

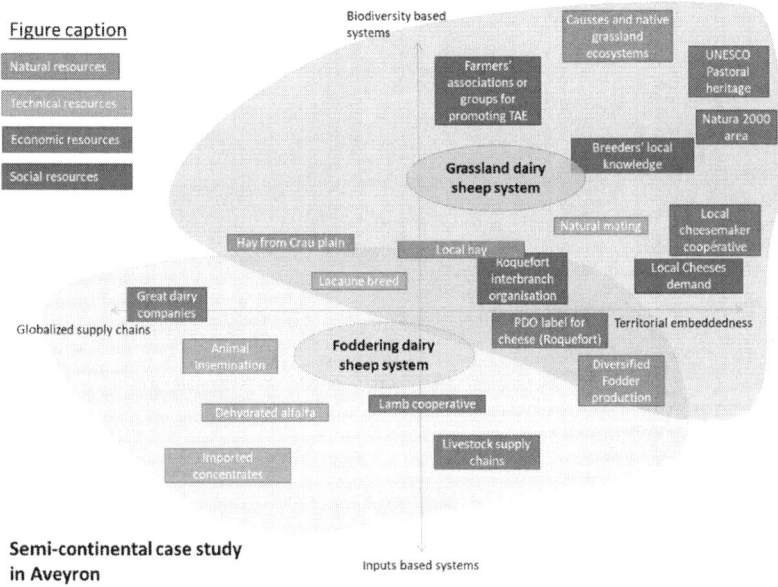

Fig. 5: Resources mobilized in the Aveyron case study.

The *Lacaune* breed is common to both systems. All farmers who participate in breed's genetic selection use animal insemination. GSS looks for other features like the robustness of animal well adapted to "extensive" breeding and grazing. In organic production, farmers use natural mating and look for compatible rams. All the farmers involved in Roquefort cheese production deliver their milk to manufacturers operating on the world market. Paradoxically, cheese production is strongly embedded in its territory, but the major actors of dairy industry sell it

Resources for agroecological transitions in livestock

worldwide. Some GSS farmers have also set up a cheese cooperative in order to enhance the value of summer and autumn milk production from animals grazing on local native grasslands.

Strong farmers' organizations and a wide range of technical support focus mainly on milk production (quality, quantity, animal feed). Farmers' groups supported by technicians design and test new agronomic practices based on local knowledge of grazing, forage cultivation or conservation agriculture. In the southern area, different actors (farmers, technicians, advisors, researchers) have recently created an "AgroEcolab"network to support local agroecological transition.

Reunion' goats are a diversification source of income with a good added value combined with a high cultural embeddedness.

In *Reunion* LFS rely on two main resources: goats ability to promote natural resources and their strong profitability on local market (Fig. 6).

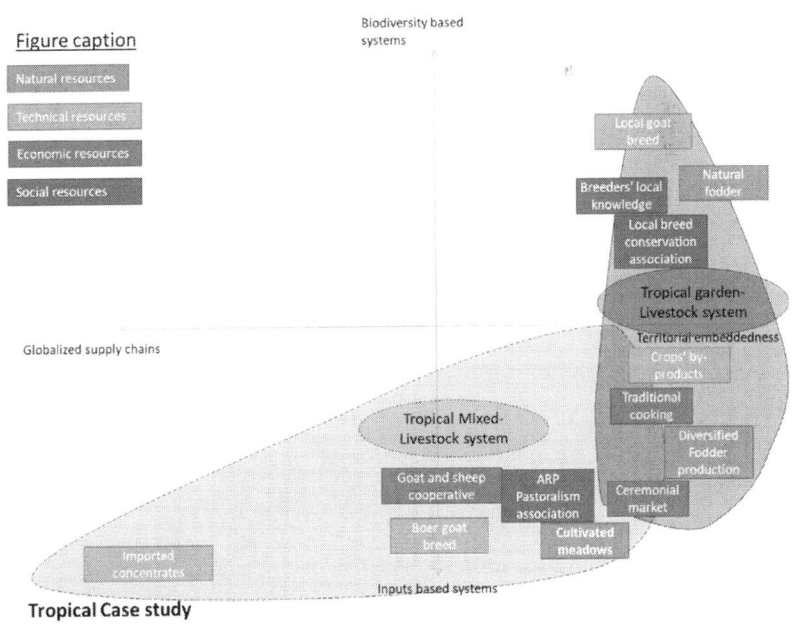

Fig. 6: Resources mobilized in the Reunion case study.

Fodder and by-products are the basis of animal feeding, but while TMS is developing a productive strategy to valorize long-term temporary grasslands; TGS opts for a multifunctional strategy of crops by-products and natural grass. Nevertheless, most farms use a large amount of concentrates from imported resources for lactating and fattening goats. The products' high profitability is due to direct sales for ritual slaughtering (for males) or short supply chains (butchers) for culling. The high costs of concentrate should not be considered only through its economic angle, but as a mean to obtain an expected conformation of animals. Moreover, cooperative market is weakly developed because of lower selling price.

There is a lack of a farmer's association which could create a real social dynamic, and this leads to a lack of technical advice. However, TMS benefits from access to services related to organized and conventional sector (Pastoralism Corporation, agriculture chamber, feed provider, financial companies, etc.). TGS relies more on a neighborhood network to access to resources like natural or cultivated fodder. Unlike cattle or sheep farming, goat farming often does not seek CAP subsidies or environmental services because the ratio between the small monetary amount and additional regulatory constraints is not attractive.

For both systems, goat genetic resources are crucial. TMS should be more oriented on imported *Boer* breed with high-quality carcasses. TGS would need a more robust breed with mother abilities and hardiness (as such *Pei, local breed*). In fact, crossbreeding has largely been used threatening the local breed. From this perspective, agroecological transition is limited by local breed's lack of a breeding scheme and the weak involvement of farmers in breeders' associations.

3.3 LFS agroecological performances in contrasted territories

LFS agroecological performances (Fig. 7) appear to be correlated with the level of biodiversity-based practices, first of all in most extensive systems based on natural grasslands (GDS, MRS, PGS). These systems reduce chemical inputs levels and improve soil fertility management at the same time, but trade-offs between the two criteria must often be sought, like FSS using herbicides to reduce soil tillage. Farms' autonomy is stronger for GSS, based on an intensification of forage production and optimal use of grass resources. Besides, farm's autonomy is hampered by limited productivity (in

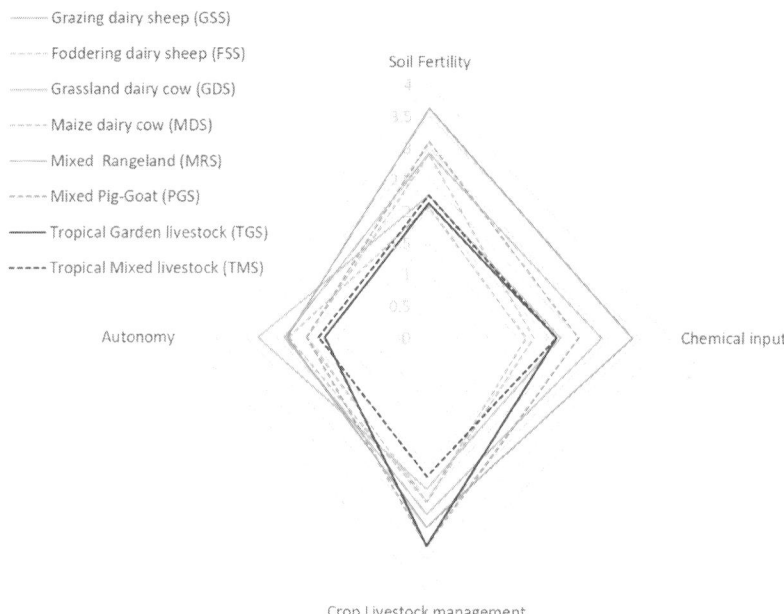

Fig. 7: Assessment of four agroecological performances for different systems.

extensive native grassland) or limited availability of land (in Reunion Island), which require to import feed and concentrates.

The crop-livestock management score arises from many combinations linked with animal and plant diversities and their integration. Dairy systems (cows and sheep) have a low level of animal diversity offsets due to a diversity of crops and forage resources. Tropical systems embody both extremes values: TMS has a very low value of plants and crops diversity while TGS is very integrated and diversified. Despite the weakness of legumes use in forage, use of manure for gardening is a cornerstone of Crop-Livestock integration.

This analysis highlights generic patterns for adapting agroecological systems to local conditions, and resource availability or constraints, leading to performance trade-offs.

Extensive systems could improve most of criteria, but with lower agricultural production. Feed self-sufficient system limits feed purchases, but not chemical inputs for forages and crops production. A highly diversified system increases integration and combination of animal and plant species without improving soil fertility and reducing the use of chemical inputs.

4. Learnings and perspectives

4.1 What resources analysis tells us about transition dynamics?

Studied LFS are based on a broad gradient from inputs-based systems in a globalized market to biodiversity-based systems deeply embedded in the territory. The position of these systems in this gradient results from individual objectives and strategies of mobilization of resources, and stakeholders interacting strategies influencing LFS territorial integration and short or long supply chains organization (Nguyen and Purseigle, 2012).

Our work illustrates how natural, technical, economic and social resources determine biodiversity-based systems' implementation and territory embeddedness. Farmers and local stakeholders build alternative sociotechnical niches based on local resources, networks and new forms of legitimacy, which justify their local relevance (Geels and Kemp, 2007).

A first learning is that the localized power relationships between stakeholders led to specific agroecological transition trajectories. The social capital (Coleman, 1988) and actors' abilities to coordinate properly also influences the baseline and agroecological systems, due to actors defining their own technical standards based on common values and objectives (e.g. the use of improved vs. rustic animal breeds). The technical and managerial strategies are therefore oriented towards certain resources: some are shared between baseline and agroecological systems, but many are specific. In our examples, local breeds, adapted to climate and local fodder resources, could be "open" resources accessible to all farmers, but are currently linked with specific networks. These include access to breed associations to procure the animals, experienced breeders, technical advisors and/or researchers to gain adequate knowledge in order to adapt management practices, commercialization channels for specific products such as labeling and branding. Success has been achieved with public support for breed maintenance (i.e. *Raïole* sheep in *Languedoc*).

The second learning is that some sets of resources become mobilized in open-ended transitions, without a determined model to follow. For instance, *Aveyron*'s LFS can follow four pathways using natural resources in different ways (Thénard et al., 2018). *Languedoc* and *Reunion* LFS diversify technical options with several species and breeds combined differently at farm level. The use of *Crau Hay* can be an opportunity to secure feed supply in case of difficult years in LFS that are usually feed self-sufficient (e.g. GSS in *Aveyron*), or to structurally provide feed (e.g. PGS in *Languedoc*). In each LFS, each breeder builds his own system according to available resources in a relatively unique combination, but with common principles.

The third learning is that specific resources can lead to determinist transitions in the sense that they exclude some types of systems (e.g. incentives from Natura 2000 policies, local breeds in *Reunion* only for agroecological niches). Other specific resources stay out of some LFS scope (e.g. in *Brittany*, the baseline system does not consider permanent grassland as productive resource and therefore does not use it). Rules such as PDO specifications or local incentives as "*Terres de sources*" determine which LFS is in or out of agroecological transition pathways, as they structure the set of accessible resources.

The final learning is that even a determinist transition can induce a more general transition. In *Brittany*, for example, organic milk cooperative started advising on grassland management, providing knowledge to all farmers around. This type of knowledge has been built within agroecological niches, but it also has been made accessible to baseline systems, authorizing hybridizations and new pathways to be developed. Flexibility can be found in these "middle way" resources, For example, performance obligation in *Bleu-Blanc-Coeur* and *Terres de sources* certifications. Even organic farming, based on an obligation of means, will not mobilize the same mix of resources according to the territory.

In conclusion, each system in transition will try to mobilize different resources, moving towards greater use of diversity and stronger territorial anchorage. In a territorial perspective, maximizing socio-agroecosystem productivity can be achieved by combining different LFS, as each one only promotes a specific part of territorial resources. The degree of local-specificity, resource availability and the sociotechnical networks will influence agroecological transitions in a more determinist or open-ended way.

4.2 LFS agroecological transitions perspectives

Our approach to define sustainable LFS is based on agroecology main principles as applied to animal production (Dumont et al., 2013). To implement these principles in order to help farmers change their practices, a first methodological development was carried out in the *Aveyron* case study (Thénard et al., 2014, 2016, 2018). The present work is an attempt to widen the methodology by testing and adapting the indicators to a set of contrasting territories. The approach allowed assessing different LFS ranks of advance in agroecological transitions, and described the set of resources mobilized, possible changes considering available resources and those to be developed.

Comparing four contrasted case studies reveals similarities and specificities of the various territories, and allows identification of pathways and options to check adequate resources and unlock agroecological transitions. A further perspective is to assess robustness and vulnerability of key resources to the hazards and possible changes.

Today, agroecology implementation is mainly considered at farm level because it affects farmers' practices and personal willingness for change. However, farms partnership offers opportunities to access and manage equipment, labor and material resources, used in agroecological transitions (Lucas et al., 2019). Despite widespread use of agroecology concepts amongst professional and political bodies, a gap remains between rhetoric and practices of implementation on farms. To assess the possible changes and transitions in farms, territorial resources inquiry seems promising, to: (i) activate resources and to favor the emergence and diffusion of innovative practices, (ii) design adequate public policies to support adequate systems, and (iii) involve consumers and citizens. In these ways, societal expectations and controversies around animal husbandry practices and animal welfare could also play a part in LFS transformations towards agroecological perspectives.

References

Affholder, F., Bessou, C., Lairez, J., Feschet, P., 2019. Assessment of Trade-Offs Between Environmental and Socio-Economic Issues in Agroecological Systems, in Côte, F.-X., Poirier-Magona, E., Perret, S., Roudier, P., Rapidel, B., Thirion, M.-C. (Eds.), *The Agroecological Transition of Agricultural Systems in the Global South*, Versailles, Quae, 219–238.

Altieri, M. A., 2002. Agroecology: The Science of Natural Resource Management for Poor Farmers in Marginal Environments, *Agriculture Ecosystems & Environment*, 93, 1, 1-24.

Bonaudo, T., Bendahan, A. B., Sabatier, R., Ryschawy, J., Bellon, S., Leger, F., Magda, D., Tichit, M., 2014. Agroecological Principles for the Redesign of Integrated Crop-Livestock Systems, *European Journal of Agronomy*, 57, 43-51.

Buclet, N., Cerceau, J., 2019. Interactions et rétroactions entre dimensions matérielle et immatérielle de systèmes communs de ressources spatialisés, une lecture par l'écologie territoriale, *Développement Durable & Territoires*, 10, 1.

Coleman, J. S., 1988. Social Capital in the Creation of Human Capital, *American Journal of Sociology*, 94, S95-S120.

Colletis, G., Pecqueur B., 2005. Révélation de ressources spécifiques et coordination située, *Économie et institutions* 6-7|2005, 9. [online] URL: http://journals.openedition.org/ei/900.

Cordier, J., Erhel, A., Pindard, A., Courleux, F., 2008. La gestion des risques en agriculture de la théorie à la mise en œuvre: éléments de réflexion pour l'action publique, *Notes et études économiques*, 30, 33-71.

Couvreur, S., Petit, T., Le Guen, R., Ben Arfa, N., Jacquerie, V., Sigwalt, A., Haimoud-Lekhal, D., Chaib, K., Defois, J., Martel, G., 2019. Déterminants techniques et sociologiques du maintien des prairies dans les élevages bovins laitiers de plaine, *Productions Animales*, 32, 3, 399-416.

De Herde, V., Maréchal, K., Baret, P. V., 2019. Lock-ins and Agency: Towards an Embedded Approach of Individual Pathways in the Walloon Dairy Sector, *Sustainability*, 11, 16, 4405.

De Vries, M., De Boer, I. J. M., 2010. Comparing Environmental Impacts for Livestock Products: A Review of Life Cycle Assessments, *Livestock Science*, 128, 1-11.

Dumont, B., Fortun-Lamothe, L., Jouven, M., Thomas, M., Tichit, M., 2013. Prospects from Agroecology and Industrial Ecology for Animal Production in the 21st Century, *Animal*, 7, 6, 1028-1043.

Elzen, B., Augustyn, A., Barbier, M., Van Mierlo, B., 2017. *AgroEcological Transitions: Changes and Breakthroughs in the Making*, Wageningen, University & Research.

Fontaine, O., Niobe, D., Shitalou, E., Fontaine, D., Choisis, J. P., 2008. Hindouisme et sacrifice de boucs à l'île de la Réunion, *Ethnozootechnie*, 85, 101–110.

Fontaine, O., Bouyssière, S., Choisis, J. P., 2010. A Typology of Goat Farming in Reunion Island Prior to the Implementation of a Breeding Scheme Adapted to the French Overseas Departments, *Advances in Animal Biosciences*, 1, 2, 510–511.

Frayssignes, J., 2001. L'ancrage territorial d'une filière fromagère d'AOC. L'exemple du système Roquefort, *Economie Rurale*, 264, 89–103.

Fuselier, M., 2019. *Studying Mixed Livestock Farming Systems in Languedoc Roussillon: Connecting Innovations, Adapting Capacities and Territory Embeddedness*, Master thesis, AgroParisTech, Paris.

Geels, F. W., Kemp, R., 2007. Dynamics in Socio-Technical Systems: Typology of Change Processes and Contrasting Case Studies, *Technology in Society*, 29, 4, 441–455.

Gerber, P. J., Steinfeld, H., Henderson, B., Mottet, A., Opio, C., Dijkman, J., Tempio, G., 2013. *Tackling Climate Change through Livestock: A Global Assessment of Emissions and Mitigation Opportunities*, Rome, Food and Agriculture Organization of the United Nations.

Gliessman, S., 2016. Transforming Food Systems with Agroecology, *Agroecology and Sustainable Food Systems*, 40, 3, 187–189.

Horlings, L. G., Marsden, T. K., 2011. Towards the Real Green Revolution? Exploring the Conceptual Dimensions of a New Ecological Modernisation of Agriculture that Could 'Feed the World', *Global Environmental Change*, 21, 2, 441–452.

Lucas, V., Gasselin, P., Van Der Ploeg, J. D., 2019. Local Inter-Farm Cooperation: A Hidden Potential for the Agroecological Transition in Northern Agricultures, *Agroecology and Sustainable Food Systems*, 43, 2, 145–179.

Madelrieux, S., Buclet, N., Lescoat, P., Moraine, M., 2017. Écologie et économie des interactions entre filières et territoires: quels concepts et cadre d'analyse? *Cahiers Agricultures*, 26, 24001.

Magne, M. A., Martin, G., Moraine, M., Ryschawy, J., Thénard, V., Triboulet, P., Choisis, J. P., 2019. An Integrated Approach to Livestock Farming Systems' Autonomy to Design and Manage Agroecological Transition at the Farm and Territorial Levels, in Bergez, J. E., Audouin, E., Therond, O. (Eds.), *Agroecological Transitions: From Theory to Practice in Local Participatory Design*. Cham, Springer, 45–68.

Martel, G., Guilbert, C., Veysset, P., Dieulot, R., Durant, D., Mischler, P., 2017. Mieux coupler cultures et élevage dans les exploitations d'herbivores conventionnelles et biologiques: une voie d'amélioration de leur durabilité? *Fourrages*, 231, 235–245.

Martin, G., Moraine, M., Ryschawy, J., Magne, M. A., Asai, M., Sarthou, J. P., Duru, M., Therond, O., 2016. Crop–Livestock Integration Beyond the Farm Level: A Review, *Agronomy for Sustainable Development*, 36, 3, 53.

Meynard, J. M., Jeuffroy, M. H., Le Bail, M., Lefèvre, A., Magrini, M. B., Michon, C., 2017. Designing Coupled Innovations for the Sustainability Transition of Agrifood Systems, *Agricultural Systems*, 157, 330–339.

Milestadt, R., Dedieu, B., Darnhofer, I., Bellon, S., 2012. Farms and Famers Facing Change: The Adaptive Approach, in Darnhofer, I., Gibbon, D., Dedieu, B. (Eds.), *Farming Systems Research into the 21st Century: The New Dynamic*. Dordrecht, Springer, 365–385.

Moraine, M., Lumbroso, S., Poux, X., 2018. Transforming Agri-Food Systems for Agroecology Development: Exploring Conditions of Success in European Case Studies, *Proceedings of the 13th International Farming Systems Association*, Chania, Greece, July 01–05.

Nguyen, G., Purseigle, F., 2012. Les exploitations agricoles à l'épreuve de la firme. L'exemple de la Camargue, *Études Rurales*, 190, 99–118.

Petit, T., Martel, G., Vertès, F., Couvreur, S., 2019a. Long-Term Maintenance of Grasslands on Dairy Farms is Associated with Redesign and Hybridisation of Practices, Motivated by Farmers' Perceptions, *Agricultural Systems*, 173, 435–448.

Petit, T., Sigwalt, A., Le Guen, R., Martel, G., Couvreur, S., 2019b. Place des prairies dans les logiques fourragères des éleveurs laitiers du Grand Ouest de la France, *Fourrages*, 239, 235–245.

Rigolot, C., Martin, G., Dedieu, B., 2019. Renforcer les capacités d'adaptation des systèmes d'élevage de ruminants: Cadres théoriques, leviers d'action et démarche d'accompagnement, *Productions Animales*, 32, 1, 1–12.

Thénard, V., Jost, J., Choisis, J. P., Magne, M. A., 2014. Applying Agroecological Principles Redesign and to Assess Dairy Sheep Farming Systems, *Options Méditerranéennes*, Série A : Séminaires Méditerranéens, 785–789.

Thénard, V., Choisis, J. P., Pages, Y., 2016. Towards Sustainable Dairy Sheep Farms Based on Self-Sufficiency: Patterns and Environmental Issues, *Options Méditerranéennes*, Série A : Séminaires Méditerranéens, 81–85.

Thénard, V., Morin, E., Frugier, J., De Boissieu, C., 2018. Quels leviers agroécologiques mobiliser pour la reconception de systèmes durables en brebis laitière in *24èmes Rencontres Recherches Ruminants*, 43–46, Paris, France, December, 05–06.

Therond, O., Duru, M., Roger-Estrade, J., Richard, G., 2017. A New Analytical Framework of Farming System and Agriculture Model Diversities. A Review, *Agronomy for Sustainable Development*, 37, 3, 21.

Wezel, A., Bellon, S., Doré, T., Francis, C., Vallod, D., David, C., 2009. Agroecology as a Science, a Movement and a Practice. A Review, *Agronomy for Sustainable Development*, 29, 4, 503–515.

Wezel, A., Brives, H., Casagrande, M., Clement, C., Dufour, A., Vandenbroucke, P., 2016. Agroecology Territories: Places for Sustainable Agricultural and Food Systems and Biodiversity Conservation, *Agroecology and Sustainable Food Systems*, 40, 2, 132–144.

The dynamics of agropastoral activities with regard to the agroecological transition

Charles-Henri Moulin, Laura Etienne, Magali Jouven, Jacques Lasseur, Martine Napoléone, Marie-Odile Nozières-Petit, Eric Vall, Arielle Vidal

1. Introduction

Agroecology emerged as a scientific discipline during the 1970s (Wezel et al., 2009). It applies ecological theory to the design and management of sustainable agroecosystems. Agroecological initiatives first focused on crops and aimed at transforming industrial agriculture, *"away from fossil fuel-based production largely for agroexport crops and biofuels"* (Altieri et al., 2012). The links between crop and animal production in the management of tropical agroecosystems in developing countries have been emphasized as an agroecological lever since the 1980s (Wezel and Soldat, 2009). For the livestock sector, recent agroecological approaches have focused on mixed crop-livestock systems (Stark et al., 2018) and examined the integration between crops and livestock at the farm level as an agroecological pathway to reduce the use of inputs in high external input agriculture or to increase outputs in low external input agriculture (Bonaudo et al., 2014). Nevertheless, few studies have considered agropastoralism (Jouven, 2016) with respect to agroecology. This is perhaps because agropastoral livestock systems use few external inputs and are therefore often considered agroecological. However, agropastoral systems are not always sustainable; they also exhibit strong dynamics. Is the agroecological transition a relevant concept to view the transformation and sustainability of this type of livestock system? Our purpose is to analyze the dynamics of agropastoral systems: What is changing in these dynamics? Who are the implied actors and what are their visions for change? To what extent are these dynamics consistent with the principles of agroecology?

By using an inductive approach, we examined the dynamics of agropastoralism in two regions: cotton areas in sub-humid West Africa (e.g., south of Mali and West Burkina Faso) and in the Mediterranean portion of the *Occitanie* region in France. In cotton areas, the reduction of rangeland areas, because of extension of crops and opportunities from milk marketing are two drivers of change for agropastoral farms. In the French Mediterranean, sources of grazing forage are diversifying and new interactions between stakeholders are being developed which provide new opportunities for agropastoral farms. We used several studies that we had previously conducted in those two regions. These studies focused on the dynamics over at least the two last decades and diversity of livestock farming systems. Data were obtained through direct interviews with selected livestock farmers. We produced typologies of farming systems with depictions of the technical operation of the systems and trajectories of farms and types of systems by using several methodologies (Cochet and Devienne, 2006; Moulin et al., 2008). Most of these studies did not focus on agroecological transition but they provided accurate data to examine changes in agropastoralism at several nested levels of organization. Then, we assessed these changes for three dimensions. The first dimension is the transition: do the changes correspond to a transition or to an adaptation? Secondly, we determined if the changes were consistent with agroecological principles. Finally, we examined the actors involved in these changes and what were their visions of change and considered the two perspectives presented in the introductory chapter: determinist or open-ended.

2. Agropastoralism and agroecology

Agropastoralism is characterized by extensive production of herbivorous livestock based on utilization of a variety of grazing areas (e.g., rangelands and cultivated lands) which is made possible by the mobility of the herds. This is a livestock-based livelihood strategy: livestock and livestock-related activities account for a large part of household incomes. It is also a way of life for which socio-cultural norms, values and local knowledge revolve around livestock (Ayantunde et al., 2011). The pastures used encompass various forms of spontaneous vegetation (e.g., steppe, savannah, forest, etc.) but also include cultivated lands (e.g. crop residues, grasslands, etc.), which justify the term agropastoral.

2.1. Agropastoralism may be in line with the principles of agroecology

Altieri (2002) identified the key ecological processes by which agroecosystems can more sustainably produce food and fibre by using fewer external inputs. From this study, Dumont et al. (2013) proposed a set of five principles for the design of sustainable animal production systems: (i) adopting management practices that aim to improve animal health, (ii) decreasing the inputs needed for production, (iii) decreasing pollution by optimizing the metabolic function of farming systems, (iv) enhancing diversity within animal production systems to strengthen their resilience and (v) preserving biological diversity in agroecosystems by adopting certain management practices.

In agropastoral systems, the use of rangelands enables decreased inputs for production (principle ii). Indeed, because areas of spontaneous vegetation are grazed, the feed resources in rangelands do not require chemical inputs or fuel for production and harvesting. Animals must move long distances to find scarce resources which are widespread over large areas and they are thus confronted with harsh environments (i.e., climate, parasites, and predators). Natural and artificial selection created animal breeds which were adapted to these harsh environments with small body sizes or disease tolerance (Mandonnet et al., 2011). Retaining these local, adapted breeds favors animal adaptation and strengthens their immune systems (principle i). Agropastoralism is also characterized by flexible and adaptive management of vegetation and of animals which do not always aim to maximize production. Grazing management allows animals to express feeding behaviours by which they compose their rations through a diversity of vegetation. Self-selection of plant secondary metabolites such as tannins reduces infestation by parasites (Villalba et al., 2010) and is an alternative to chemical drugs. Flexible culling management causes a diversity of reproductive trajectories by females with unproductive periods for some females with better long-term viability of the herd (Tichit et al., 2004). Taking advantage of vegetation and animal diversity with adaptive management strategies enables increased resilience of the system (principles iv and i). Animals also graze forage resources on grasslands (permanent or cultivated) and crop residues. Through their daily mobility, animals link various areas in agropastoral landscapes. Animal manure provides a key energy source for soil microorganisms. They also provide nutrients for crops with transfer from rangelands which are grazed during the day to small plots of cultivated land where animals are kept during the night. The horizontal nutrient fluxes from large rangeland are not so important by unit area and compensate by their vertical fluxes of nutrients from the deep horizons to topsoil thanks to the presence of shrubs and trees on rangelands.

Agropastoral communities often develop collective rules for management of common resources: access to rangelands and rights of commonage to crop residues after a given date, such as in done in West Africa (Grillot et al., 2018). Thus, agropastoral systems allow better regulation of biogeochemical cycles through spatial and temporal interactions among different farms (Gliessman, 2006) and decreased pollution thanks to the metabolic functions of farming systems (principle iii). Finally, agropastoral farmers recognize and use plant and animal diversity which are valued as productive assets and not only as a tradition (principle v). Farmers and shepherds develop local knowledge of the ecology and behavior of plants and animals (Bollig and Schulte, 1999) and rely not only on their skills but also on the adaptive abilities of animals and plants. Animal skills are shaped through selection of individuals by the farmer and through the learning process of each individual animal which takes place in the pasture from a young age through individual experiences and observations of peer behaviors (Meuret and Provenza, 2015).

2.2. Agropastoral systems are changing

Agropastoral systems are often seen as traditional systems which stand apart from the modernization movement of agriculture. Nevertheless, agropastoralism has evolved in many parts of the world under the effects of several drivers. Different public policies such as settling of nomadic peoples or prohibiting access to protected natural areas has decreased the mobility of agropastoral families who were forced to find new combinations of resources to feed their herds. Changes in land use are another strong driver of change. In developed countries, industrialization of the food system has led to specialization of farms. In less-favored areas, farmers abandoned food crops and converted arable lands to grasslands to feed animals in a livestock specialization environment. Workforce diminution and moto-mechanization of forage production

Agropastoral dynamics and agroecological transition

on grasslands led to decreased use of grazing on rangelands (Aubron et al., 2016). This decreased use of rangelands has resulted in the advance of shrubs and trees on rangelands. In developing countries, increased rural populations led to expansion of cultivated areas at the expense of rangelands. Animal numbers have also increased and agropastoral farmers must find new resources to feed them. In northern Africa for instance (Bourbouze, 2000), sheep flocks raised on the steppe rangelands are also fed with cereal grains. This enables maintaining a high number of grazing animals than the local land can feed which raises complex issues regarding degradation of soils and vegetal communities.

We found evidence in the scientific literature that agropastoral systems may be consistent with agroecological principles. This does not mean that all agropastoral systems rely on agroecological principles and are sustainable. In developed countries, rangelands can be still owned by livestock farmers but with little utilization. In developing countries when the ratio of rangelands to arable lands decreases dramatically, the remaining rangelands may be overgrazed with a loss of biological diversity. Carbon and nutrient flows between various areas of the landscape are no longer effective in maintaining soil fertility on arable land and chemical fertilizers are used. Thus, the presence of rangelands in farm assets is not sufficient to guarantee the agroecological characteristics of the system.

3. A framework to analyze livestock dynamics in agropastoral regions

In the dynamics that we studied, we identified changes at four levels. First, we considered the logic of herd management and of feed resources at the livestock system level. Following Francis et al. (2003), who redefined agroecology as *"the integrative study of the ecology of the entire food system, encompassing ecological, economic and social dimensions,"* we also considered three upper levels: farms and livelihoods of farm families; diversity of farm types and the relationships between these types at the territory level; and local food systems which supply food to urban and rural families in the territory. We then qualified these changes. If the dynamic did not imply a change of the structure and feedback mechanisms of the system, we classified it as an adaptation. In contrast, when a dynamic implied changes in the structure (new entities to manage) and management of the entities involved for at least one level, we classified this dynamic as a transition. Finally, we assessed these changes with regard to agroecological principles.

Before discussing our studies in the two regions, Tab. 1 synthesizes the analysis results. We identified five dynamics. Two of these dynamics rely on the intensification of production on animals (CA1) or grasslands (FM1) with external inputs. These changes are not consistent with agroecological principles. These dynamics are conventional transitions of intensification with external inputs and an exit from agropastoral and agroecological logic. In the third dynamic (FM2), farmers still rely on rangelands in combination with a few grasslands. They maintain an agropastoral logic to feed their herds which is based on agroecological principles. However, they enlarge their herds to maintain income and seek to access more rangelands by negotiating with a diversity of landowners. This dynamic is an adaptation of agropastoral farms to changes in their socio-economic environments. In this adaptive dynamic, the agroecological and agropastoral logic were preserved at the livestock system level. The last two dynamics (CA2 and FM3) are agroecological transitions. For the CA2 dynamic, farmers change their management for parts of the herds. From an agropastoral management perspective and based on grazing and mobility of the herd, feeding a batch of lactating cows moves to a crop-livestock integration. This new management approach still relies on agroecological principles, those classically described for crop-livestock integration, which are different from those of agropastoral management. This is a transition which is based on agroecological principles at the livestock system level first but also at the farm and local food system levels. Finally, the FM3 dynamic corresponds to new arrangements between specialized farmers (crop or livestock) and other territorial actors. These new arrangements enable agropastoral farmers to access new feed resources from cultivated lands managed by crop farmers. This dynamic is a transition at the farm and territory levels. These changes are consistent with agroecological principles and still consider the relevance of crop-livestock integration and recycling in the landscape. At the livestock system level, agropastoral farmers retain the same agropastoral logic and its agroecological principles. Thus, we classified this transition as agroecological because of the changes at the territory level which were consistent with agroecological principles and also because it enabled keeping agropastoral farms in the territory.

Agropastoral dynamics and agroecological transition 231

Dynamics	CA1	CA2	FM1	FM2	FM3
Herd and feed resources / Livestock system	Animal intensification with exotic cows fed with purchased feeds	Crop-livestock integration for the batch of dairy cows	Fodder intensification by moto-mechanisation and chemical fertilizers	*Keeping the same agropastoral logic of management*	*Keeping the same agropastoral logic of management*
Farm - Family	No more use of rangelands	Improvement of rural family food security (milk self-consumption, incomes)	Herd enlargement Increase of grasslands area Abandonment of rangelands	Herd enlargement Access to new rangelands	Access to feed resources from crop lands in other farms
Territory / Agrarian system				Arrangement to access public lands Collective pastoralism	Exchanges between specialized farms, mediated with territorial actors
Territory / Local food system	Providing local urban markets with local milk Use of local by-products				
Qualification Dynamic Transition level Agroecology Agropastoralism	Transition Livestock syst. Exit Exit	Transition Livestock syst. Keeping on Partial exit	Transition Livestock syst. Exit Exit	Adaptation - Keeping on Keeping on	Transition Farm / Territory Keeping on Keeping on

CA: cotton areas of West Africa / FM: French Mediterranean (*Occitanie* region)
White text with black background: changes that are not consistent with agroecological principles
Black text with grey background: changes that are consistent with agroecological principles

Tab. 1: Qualification of the five dynamics of livestock activities in two agropastoral regions.
CA: cotton areas of West Africa/**FM**: French Mediterranean (*Occitanie* region)
White text with black background: changes that are not consistent with agroecological principles
Black text with grey background: changes that are consistent with agroecological principles

4. In West African cotton areas, a dairy intensification pathway as a transition between two forms of agroecological livestock systems

In the cotton areas of South Mali and West Burkina-Faso, the climate is characterized by a short rainy season (four months). The agrarian systems are based on rain-fed cereal cultivation for food consumption (e.g., maize and sorghum) with some legumes (e.g., peanuts) and cotton. Livestock is also present with herders arriving from the Sahel after the severe droughts of the 1970s and 1980s. At the same time, cotton companies supported the development of draught cattle for cotton cultivation.

Then, families who mostly conducted cropping activities also developed cattle herds. In all farms, livestock graze on rangelands and crop residues which are mainly cereal straws. If the herd mobility supports transfer of organic matter between rangelands and croplands, chemical fertilizers (and pesticides) are also used for cotton.

This agropastoral system relies on agroecological principles. It uses only non-edible resources to feed the animals (e.g., forage from rangelands and crop residues) and valorizes local feed resources with local knowledge of rangeland use. It does not use fuel to produce and harvest feeds. The animals are adapted to scarce feeding resources (i.e., small body sizes plus use and reconstitution of body reserves) and are tolerant to diseases, especially trypanosomiasis. This system participates in the global metabolism of the agrarian system, links rangelands and croplands, and enhances soil fertility. As stressed by the FAO (2016), agroecology *"has been carried out by African farmers and pastoralists for millennia. Thus, while often not explicitly termed 'Agroecology,' many actors and initiatives exist within sub-Saharan Africa that builds on agroecological principles."* The development of these livestock systems in cotton areas since the 1970s is one example.

Milk is another product supplied by cattle. Herders milk some cows during the rainy season when they are able to graze fresh grasses on rangelands. Daily milk production per cow is very low: 1.2–1.5 litres (Sib et al., 2018). In West African countries, after the failure of industrialization of milk production and processing from the 1950s to 1980s, consumption of dairy products is mainly based on importation of milk powder. A new model of mini-dairies emerged in the 1990s to provide local milk to the growing urban population. This is an opportunity for livestock farmers but milk delivery to mini-dairies during the dry season is a real issue. Another issue is the extension of the crop areas at the expense of the rangelands. Intensification of milk production is therefore a challenge for agropastoral farms. Could this dairy intensification be carried out by maintaining a logic of production based on agroecological principles?

4.1. Emergence of two pathways for dairy intensification

For over two decades around the cities of Sikasso (Mali) and Bobo-Dioulasso (Burkina-Faso), we observed two main pathways of dairy intensification (Coulibaly et al., 2007; Vidal et al., 2020). The first pathway (CA1) is a zero-grazing system in a conventional transition of intensification. The second pathway (CA2) is a transition from agropastoralism to crop-livestock integration for a batch of lactating cows.

In the first intensification pathway, farmers developed a zero-grazing system and kept their cows in barns or yards. They fed them with purchased feeds (e.g., cut and carry fresh grass, hays, and concentrates). This enabled actual dairy intensification at the cow level (e.g., 800–2,000 litres of milk per lactation) and provided urban markets with local milk especially during dry seasons. But this approach no longer satisfied agroecological principles for milk production. Indeed, to valorize the expensive feed resources, farmers used artificial insemination of local cows with imported semen from exotic dairy bulls (e.g., *Montbéliarde*, Brown Swiss or Holstein). They used chemical drugs to prevent trypanosomiasis because of the sensitivity of the crossbred animals to this disease. Because feeds were purchased, there were longer any direct links between livestock operations and the natural conditions and cycles of the ecosystems delivering these feeds. The feeding logic is now based on adapting to market feed prices and the availability of cash to pay for feeds. Few farmers are currently engaged in this dynamic.

In the second pathway, farmers changed their management of batches of lactating cows to produce milk during the dry season. The feeding system for these batches was still partly based on grazing but stored feeds were also provided. Farmers collected straw from their plots and stored it in shelters. So, this straw had better nutritional value compared to that left in the field. In some cases, they cropped some fodder legumes (e.g., *Mucuna spp.* and *Vigna unguiculata*). They also used concentrates which came from household activities (cereal brans) but many concentrates were purchased such as cotton seed cakes. Farmers still mainly rely on local breeds but they also crossbreed their animals with West African zebu breeds which have a better dairy potential (e.g., *Goudali* from Niger). Some farmers also tried artificial insemination with exotic dairy breeds. Total milk production increased from an annual production of 350 litres per cow up to 600 litres (Sib et al., 2018) but above all, this production was better distributed during the year, particularly in the dry season.

4.2. A pathway of dairy intensification consistent with agroecological principles at three levels

At the livestock system level, the herder divides his herd in two batches, namely the lactating cows that he milks and the remaining animals. He manages the batch of cows using a new logic of crop-livestock integration which is consistent with other agroecological principles. Indeed, there is a classic cycle of matter from cereal straws and legume hay to the animals and, in return, from animals producing manure, mixing part of the straw and animal excreta, to cultivated plots. These new practices require more transportation of matter but are still based on human and animal work and are not fuel-based. No fertilizers are used for the legume fodders. There is also higher diversity of animal genetics with mixing of well-adapted local cattle with some crossbred animals with higher dairy potential. There is also a higher diversity of resources to feed the herd. The diversification of cultivated resources enhances the resilience of the livestock operation in a context where rangeland areas are decreasing.

At the farm level, this intensification pathway produces more milk and increases self-consumption of dairy products by families. The sale of milk to mini-dairies also increases cash incomes. Thus, the development of dairy activity is a way to diversify family livelihoods. It does not take place at the expense of family food security as cereal production does not decrease: the fodder crop areas are not very important and some may be multi-purpose (fodder or grains, such as *Vigna unguiculata*). Thus, this dairy diversification enhances family food security and reduces its vulnerability.

At the local food system level, two elements reinforce the agroecological characteristics of this pathway of dairy intensification. First, this intensification increases the share of local milk which covers urban milk consumption and could therefore decrease imports of milk powder. Thus, it enhances the food sovereignty of the region (Altieri et al., 2012). Second, the purchased concentrates are locally produced: production of brans from the transformation of grains for human food and of cotton seed cake from the industrial transformation of cotton grains. Those local by-products are utilized through livestock to provide milk and meat to the local food system. Thus, there is local recycling at the food system level. This recycling is also relevant in terms of local food security because those concentrates are not edible. Indeed, in this local food system, livestock do not compete for human food which is not the case when grains are used as concentrates.

4.3. Public actors with a determinist perspective of dairy intensification

Food security and economic development are targets of public actors who support dairy intensification such as state livestock services or research institutions. They advocate and disseminate the technical levers of intensification: artificial insemination, fodder crops, and concentrates without using agroecological thinking. They carry a determinist vision of dairy intensification with causal relationships between adoption of techniques and increased milk production. Nevertheless, since the end of the major development projects and implementation of the programs of structural adjustment in the 1980s and 1990s, public actors have few means to disseminate these technical levers. Therefore, farmers who engage in dairy intensification gradually modify their livestock management by adopting new practices which are adapted to their possibilities and seize opportunities from various initiatives (e.g., public services, NGOs, and private operators of milk chains who may develop other visions of change). In that sense, the trajectories of farms are rather opportunistic. Thus, the changes made in the agroecological intensification pathway are many and varied.

The different actors who support farmers share the same interest in milk production for the local food system. There is also a consensus regarding the need for changes in the ways of producing milk by increasing the milk yield per cow during the dry season. These actors have various visions for the means to obtain such increased milk yields. It is a real challenge to strengthen a dairy intensification effort in an agroecological way which is adapted to various local conditions to face the development of conventional intensification for whom the no-grazing system is one model. The recent development of innovation platforms in Burkina-Faso with participatory approaches is somewhat relevant to allow confronting these visions and develop a holistic approach to the complexity of local milk value chains. Linking the actors of local milk value chains enables sharing the visions of change of various stakeholders and helps to test new practices at the livestock system level (such as forage production) which are adapted to local contexts and to the diversity of farm types.

5. In the French Mediterranean, transitions at the farm and territory levels enable maintaining the agroecological logic of agropastoral livestock systems

We distinguish two main landscapes in the Mediterranean part of the *Occitanie* region. First, the coastal plain gathers the main urban poles, transport facilities and tourist equipment. Historically, the agrarian system was based on integrated mixed crop-livestock systems. Livestock activities used both rangelands and crop residues or vineyard undergrowth. Livestock contributed to clearing grass/weeds and to retaining soil fertility. During the modernization of agriculture, irrigation facilities and favorable climatic and soil conditions led to specialization in high-value crops from vineyards, orchards, and market gardens and to a strong decrease in livestock numbers. The interstitial rangelands were not used anymore. The recent evolution of the wine economy has also led to the abandonment of lands that have become wastelands.

In the mountainous hinterlands, agriculture modernization has led to livestock specialization and the abandonment of many rangelands. Forests are now the main land use. At higher altitudes, alpine meadows are used in the summer by transhumant systems. At lower altitudes in the valleys, small surface areas of grasslands constitute a crucial resource for livestock activity. At intermediate altitudes, encroached rangelands are also present in a mosaic with forests.

In these two landscapes, utilization of spontaneous vegetation is a very important issue to maintain landscape mosaics, conserve biodiversity which is linked to open landscapes, prevent forest fires, and preserve cultural landscapes which are valued for their tourism and recreational activities. Livestock farming systems are expected to make sustainable use of vegetation to fulfil such objectives. Maintaining farm families in the hinterlands is also expected to contribute to local economic development. In this context, we identify three dynamics over the past decades.

5.1. Pathway of forage intensification: From agropastoral to cultivated grass-based livestock systems

The first dynamic (FM1) is the development of livestock activities by native families in the hinterlands who accumulated favorable lands from families leaving agricultural activities and is based on grasslands.

Agropastoral dynamics and agroecological transition 237

They abandoned the use of rangelands. These farms use more chemical inputs (fertilizer for the grasslands) and fuel for grass cultivation and harvesting (moto-mechanization). For cultivated grasslands, few species are used at the plot and at farm levels. The farms are self-sufficient for forage and a portion of concentrates when cereals are cultivated in rotation with grasses. However, feeding animals only from grasslands means that there is no longer any connection between different compartments of the landscape. Such livestock systems are no longer agropastoral ones. Dairy systems (e.g. ewes and goats) are often in this dynamic but so are some suckling cattle systems. This new technical logic of livestock farming is a transition but is not an agroecological transition.

5.2. A dynamic of adaptation of agropastoral farms

In a second dynamic (FM2), farmers have maintained an agropastoral logic at the livestock system level with adaptations to the changing socio-economic environment. These farms are located in the hinterlands and have retained a feeding system which is based on a combination of several resources including rangelands, permanent grasslands and cultivated lands (e.g., fodder and cereals). This dynamic concerns mostly meat livestock systems (sheep and cattle). This specialization in meat operations has led to increases in animal numbers to maintain farm incomes. Searching for new opportunities to access cheap feeding resources is essential to accommodate this increase in animal numbers. Farmers must negotiate with various (and sometimes many) private or public landowners. The French Mountain Law of 1972 renewed the development of collective pastoralism in the mountains mainly to manage access to high-altitude meadows. The development of agro-environmental measures in the second pillar of the CAP also increases grazing in high natural value areas. Those measures maintain the capacity of agropastoral farmers to continue their access to rangelands.

In these agropastoral systems, fattening young animals is difficult without feed purchases (concentrates or high-quality forage). Thus, these systems have changed and have utilized several combinations of feeding resources and types of marketed animals along the trajectory of the farms but they retained an agropastoral logic. They still rely on the same agroecological principles, mainly the use of few inputs, due to grazing of spontaneous vegetation by females and the links between rangelands and croplands through animal mobility and manure use.

5.3. An agroecological transition at the territory level for agropastoral farms accessing feeds on cultivated lands

The third dynamic (FM3) often corresponds to the creation of new farms in the hinterlands and on coastal plains. New farmers begin pastoral livestock operations based on rangelands which are easier to access than cultivated lands. They then seek access to feed resources on cultivated lands and develop exchanges with specialized crop farms. This trend is connected to the evolution of the technical operations in the vineyards or orchards with reintroduction of grasses between rows of vines or trees and societal pressure to reduce herbicide use. To take advantage of these new resources, breeders do not need to make in-depth changes to the agroecological logic of their feeding and grazing systems. Nevertheless, to access new feed resources on cultivated lands, they have developed new types of crop-livestock integration at the territory level and have developed exchanges between specialized farms. We distinguish several types of these relationships.

In a first type of relationship, crop farmers and livestock farmers make arrangements based on neighbouring and historical relationships inside rural communities. Breeders graze crop residues after harvesting or legume pastures used in cereal-crop rotations. This form of interpersonal arrangement had been declining due to the regional specialization of agriculture and breakdown of rural communities. However, new opportunities continue to develop. In the coastal plain areas and especially near urban areas, interest in local pasture-based livestock farms has been renewed due to considerations of the potential contribution of grazing to management of natural areas but also due to reductions in herbicide and fertilizer use in cultivated areas. Thus, various municipalities and communities of municipalities have attempted to create favourable conditions for the return of agro-pastoral farms in lowland areas.

In a second type of relationship, crop-livestock integration corresponds to distant production of winter forage on vineyard wastelands in coastal peri-urban areas. For instance, cooperative action enabled connection of a group of farmers with local stakeholders in a peri-urban coastal area to initiate actions based on mid-term win-win interests. Such cooperation secures farm feeding systems in the mountains (new farms setting up mostly on rangelands and without access to grasslands in the mountains) and provides environmental services in the plain (Napoléone et al., 2019). Continuity of this action relies on evidence that livestock farmer activities in coastal zones are more than a service delivery for which others could be more efficient. It should maintain equilibrium between the two poles of livestock farming (e.g., hinterlands and coastal plain territories).

All of these new arrangements which link agropastoral farmers with other stakeholders (at several geographical levels from the local vicinity to long distances) to access new feed resources are changes that we consider to be a transition at the farm and territory levels because of the management of new types of relationships. These arrangements allow consistency with agroecological principles at the livestock system level for livestock farms. They also lead to the development of agroecological practices for specialized crop farms (reduced use of herbicides and fertilizers and increased soil organic matter content). The relevant interest of crop-livestock integration now operates between farms and not at the level of a mixed system developed on one farm. These new pastoral resources are devoted to multiple uses and are thus good examples of land sharing. Obviously, the diversity of objectives and issues associated with the various users complicates management of such land and highlights the need for new tools, better communication between stakeholders, and new skills and knowledge for farmers.

5.4. New territorial actors enable the emergence of new coordination mechanisms

Before the 1990s, access to pastoral resources for farmers was negotiated through interpersonal relationships (farmer-private landowner) and through national sectoral policies (agriculture, forest, and environment) to access public land. The institutional actors held a rather determinist view of change. Indeed, definition and implementation of these policies had particular objectives for the rangelands with a planned vision; for instance, for the expected vegetation states on those lands to maintain biodiversity. In this context, agropastoral farmers had to design sustainable feeding systems by combining a variety of resources with specific expectations from several sectoral policies which were not coordinated. This could cause contradictions that agropastoral farmers had to solve by themselves and thus generated a great variety of agropastoral systems.

The agroecological transition of the third dynamic (FM3) at the farm and territory levels is the result of the actions of territorial actors who participated in new coordination mechanisms which enabled agropastoral farmers to access a more diversified range of pastoral resources (in rangelands or arable lands). New public actors have emerged since the decentralization policy in France which started in the 1960s and also since the 1980s as Regional Natural Parks or municipality communities. But there are also various local associations from economic sectors or civil society which have various stakes or initiatives such as building links between farmers and consumers through collective organization for direct selling. In this context, with the confrontation of various visions and objectives carried by this diverse group of actors, the coordination mechanisms are rather open-ended. Agropastoral farmers are still faced with the challenge of designing livestock systems which consider expectations of multiple actors but interactions with these territorial actors can facilitate this work by making it possible to negotiate room for manoeuvring.

6. Conclusion

Various drivers (e.g., demography, climate change, and socio-economic conditions and policies) have resulted in changes in agropastoral farms. Conventional intensification with high external inputs is a strong pathway for livestock farms which exit from agropastoralism and its agroecological logic. We encountered this dynamic in the two regions studied and in many other agropastoral regions. Other farms retained agropastoral systems, but they were not static and had also evolved. This evolution could be an adaptation to external drivers without a change of the technical logic at the livestock system or farm levels. Finally, we explored two original types of agroecological transitions. In the first type, the technical logic changed from agropastoral to crop-livestock integration and was based on other agroecological principles. The confrontation between visions the actors of the value chains, which included public services and researchers, seemed to be a good way to develop open-ended perspectives and to provide a future for such agroecological pathways. The second type of transition is based on crop-livestock integration at the territory level and mobilizes specialized farmers (livestock or crop productions) and territorial actors. This transition implies agroecological changes for crop farmers for the purpose of reducing pesticide use (for instance). In contrast, livestock farmers retain the same agropastoral and agroecological logic. These two original pathways show that it is relevant to consider that an agroecological transition is not only the ecologization of intensive systems. When facing changes in context, agropastoral farms may evolve toward livestock systems that are less sustainable. It is a significant challenge to view the dynamics of agropastoral systems in terms of agroecological transitions and to contribute to designing new systems or territorial arrangements that allow the sustainability of agropastoral farms. In the context of systems with low inputs and low productivity and with the issue of food security, the challenge is to design intensification pathways while maintaining a logic of production which is based on knowledge of functioning ecosystems. In other contexts, when food security is no longer an issue at the territory level, the capacity of agropastoral systems to ensure several functions for farmers and territorial actors is another challenge for the design.

To analyze the dynamics of agropastoral systems with regard to agroecological transitions, we conducted multi-scalar analysis. By its very nature, agropastoral farming systems necessarily link several stakeholders with farm management, value chains and natural resources which reinforces the need for such multi-scalar analysis. This is also the case for mixed crop-livestock systems (Moraine et al., 2016) and for initiatives which consider agroecological transitions at landscape scales (Gascuel and Magda, 2015). We wish to focus here on the three original points of our study. First, it is useful to analyze what is changing – or not – at several levels. For the French Mediterranean case, the changes in relationships between specialized farms and changes in territorial actors' arrangements reveal a transition. Second, to classify the transition in an agroecological perspective, it is useful to have a precise analysis of the changes at the level concerned to better understand the nature of the changes. For the West African cotton areas, changes were implemented at the level of a batch of lactating cows. It was at that level that we pointed the movement of an agroecological logic to another. Finally, the multi-level analysis allowed to have a better qualification of some practices. The use of concentrates for lactating cows in cotton areas is compliant with agroecological principles of recycling, when we consider the operation of the whole local food system.

References

Altieri, M. A., 2002. Agroecological Principles and Strategies for Sustainable Agriculture, in Uphoff, N. T. (Ed.), *Agroecological Innovations: Increasing Food Production with Participatory Development*, London, Earthscan Publication Ltd, 40–46.

Altieri, M. A., Funes-Monzote, F. R., Petersen, P., 2012. Agroecologically Efficient Agricultural Systems for Smallholder Farmers: Contributions to Food Sovereignty, *Agronomy for Sustainable Development*, 32. URL: https://doi.org/10.1007/s13593-011-0065-6.

Aubron, C., Noel, L., Lasseur, J., 2016. Labor as a Driver of Changes in Herd Feeding Patterns: Evidence from a Diachronic Approach in Mediterranean France and Lessons for Agroecology, *Ecological Economics*, 127, 68–79.

Ayantunde, A. A., de Leeuw, J., Turner, M. D., Said, M., 2011. Challenges of Assessing the Sustainability of (Agro)-Pastoral Systems, *Livestock Science*, 139, 30–43.

Bollig, M., Schulte, A., 1999. Environmental Change and Pastoral Perceptions: Degradation and Indigenous Knowledge in Two African Pastoral Communities, *Human Ecology*, 27, 493–514.

Bonaudo, T., Bendahan, B. A., Sabatier, R., Ryschawy, J., Bellon, S., Léger, F., Magda, D., Tichit, M., 2014. Agroecological Principles for the Redesign of Integrated Crop-Livestock Systems, European *Journal Agronomy*, 57, 43–51.

Bourbouze, A., 2000. Pastoralisme en Afrique du Nord: la révolution silencieuse, *Fourrages*, 161, 3–21.

Cochet, H., Devienne, S., 2006. Fonctionnement et performances économiques des systèmes de production: une démarche à l'échelle régionale, *Cahiers d'Agricultures*, 15, 578–583.

Coulibaly, D., Moulin, C. H., Poccard-Chappuis, R., Morin, G., Corniaux, C., 2007. Evolution des stratégies d'alimentation des élevages bovins dans le bassin d'approvisionnement en lait de la ville de Sikasso au Mali, *Revue d'Elevage et de Médecine Vétérinaire des Pays Tropicaux*, 60, 103–111.

Dumont, B., Fortun-Lamothe, L., Jouven, M., Thomas, M., Tichit, M., 2013. Prospects from Agroecology and Industrial Ecology for Animal Production in the 21st Century, *Animal*, 7, 1028–1043.

FAO, 2016. Report of the Regional Meeting on Agroecology in Sub-Saharan Africa. Dakar, Senegal, 5–6 November 2015, Rome, FAO.

Francis, C., Lieblein G., Gliessman, S., Breland, T. A., Creamer, N., Harwood, Salomonsson, L., Helenius, J., Rickerl, D., Salvador, R., Wiedenhoeft, M., Simmons, S., Allen, P., Altieri, M., Flora, C., Poincelot, R., 2003. Agroecology: The Ecology of Food Systems, *Journal of Sustainable Agriculture*, 22, 99–118.

Gascuel, C., Magda, D., 2015. Gérer les paysages et les territoires pour la transition agroécologique, *INRA Innovations Agronomiques*, 43, 95–106.

Gliessman, S. R., 2006. *Agroecology: The Ecology of Sustainable Food Systems*, Boca Raton, CRC Press.

Grillot, M., Guerrin, F., Gaudou, B., Masse, D., Vayssières, J., 2018. Multi-Level Analysis of Nutrient Cycling Within Agro-Sylvo-Pastoral Landscapes in West Africa Using an Agent-Based Model, *Environmental Modelling & Software*, 107, 267–280.

Jouven, M., (dir.), 2016. *L'Agroécologie, du nouveau pour le pastoralisme?* Avignon, Association Française de pastoralisme et Cardère éditeur.

Mandonnet, N., Tillard, E., Faye, B., Collin, A., Gourdine, J. L., Naves, M., Bastianelli, D., Tixier-Boichard, M., Renaudeau, D., 2011. Adaptation des animaux d'élevage aux multiples contraintes de régions chaudes, *INRA Productions Animales*, 24, 41–64.

Meuret, M., Provenza, F. D., 2015. When Art and Science Meet: Integrating Experiential Knowledge of Herders with Science of Foraging Behavior for Managing Rangelands, *Rangeland Ecology and Management*, 68, 1–17.

Moraine, M., Duru, M., Therond, O., 2016. A Social-Ecological Framework for Analyzing and Designing Integrated Crop–Livestock Systems from Farm to Territory Levels, *Renewable Agriculture and Food Systems*, 32, 43–56.

Moulin, C. H., Ingrand, S., Lasseur, J., Madelrieux, S., Napoléone, M., Pluvinage, J., Thénard, V., 2008. Comprendre et analyser les changements d'organisation et de conduite de l'élevage dans un ensemble d'exploitations: propositions méthodologiques, in Dedieu, B., Chia, E., Leclerc, B., Moulin, C. H., Tichit, M. (Eds.), *L'élevage en mouvement. Flexibilité et adaptation des exploitations d'herbivores*, Paris, QUAE.

Napoléone, M., Gravas, O., Rouquette, A., Cittadini, R., Campoy, E., 2019. L'élevage et les friches au cœur de complémentarités entre littoral périurbain et arrière-pays. L'exemple du projet Fricato en Pyrénées Orientales, *Innovations Agronomiques*, 72, 107–119.

Sib, O., Bougouma-Yameogo, V. M. C., Blanchard, M., Gonzalez-Garcia, E., Vall, E., 2018. Production laitière à l'ouest du Burkina Faso dans un contexte d'émergence de laiteries : diversité des pratiques d'élevage et propositions d'amélioration, *Revue d'Elevage et de Médecine Vétérinaire des Pays Tropicaux*, 70, 81–91.

Stark, F., González García, E., Navegantes, L., Miranda, T., Poccard-Chapuis, R., Archimède, H., Moulin, C. H., 2018. Crop-Livestock Integration Determines the Agroecological Performance of Mixed Farming Systems in Latino-Caribbean Farms, *Agronomy for Sustainable Development*, 38. URL: https://doi.org/10.1007/s13593-017-0479-x.

Tichit, M., Ingrand, S., Moulin, C. H., Cournut, S., Lasseur, J., Dedieu, B., 2004. Analyse de la diversité des trajectoires productives des femelles reproductrices: intérêts pour la modélisation du fonctionnement du troupeau en élevage allaitant, INRA Productions *Animales*, 17, 123–132.

Vidal, A., Lurette, A., Nozières-Petit, M. O., Vall, E., Moulin, C. H., 2020. The Emergence of Agroecological Practices on Agropastoral Dairy Farms in the Face of Changing Demand from Dairies, *Biotechnology, Agronomy, Society and Environment*, 24, 163–183.

Villalba, J. J., Provenza, F. D., Hall, J. O., Lisonbee, L. D., 2010. Selection of Tannins by Sheep in Response to Gastrointestinal Nematode Infection, *Journal of Animal Science*, 88, 2189–2198.

Wezel, A., Soldat, V., 2009. A Quantitative and Qualitative Historical Analysis of the Scientific Discipline of Agroecology, *International Journal of Agricultural Sustainaibility*, 7, 3–18.

Wezel, A., Bellon, S., Doré, T., Francis, C., Vallod, D., David, C., 2009. Agroecology as a Science, a Movement and a Practice. A Review, *Agronomy for Sustainable Development*, 29, 503–515.

What models of justice for the agroecological transition? The normative backdrops of the transition

Pierre M. Stassart, Antoinette M. Dumont,
Corentin Hecquet, Stephanie Klaedtke,
Camille Lacombe, Matthieu de Nanteuil

1. Introduction

There is not a singular model of agroecological transition. And if this is the case, it is not only because the processes of the transition are always situated, complex, uncertain and undetermined. Behind these difficulties hide issues of another order. The thesis defended in this chapter is that they reflect on the axiological and normative bases of the transition:

- Firstly, axiological. We cannot reduce the actors of the transition to simple strategists seeking only to defend their interests or pure idealists striving for values that are disconnected from reality. They certainly have values, but these values can come into tension with one another. In practice, the actors of the transition are often confronted with conflicts of value that, if not addressed directly, can generate inhibition and suffering.
- Next, normative. It is worth leaving behind the opposition between the absence of normativity (it would suffice to support all participatory initiatives) and authoritarian normativity (it would suffice to enforce general principals). In practice, neither of these paths seems able to provide to actors on the ground with concrete tools allowing them to overcome the conflicts of values that they encounter, without renouncing their deeper motivations.

In this context, the agroecological transition may be threatened from within, due to the lack of approaches or methodologies able to meet these challenges. Therefore, how do we proceed?

To answer this question, we carry out our reflection in three steps. First, we return to the relationship between "transition" and "transformation" and, more broadly, to the meaning of an open-ended and non-relativist agroecological transition process. To this end, we will identify and go through three "normative stages" for agroecology: the first corresponds to a statement of general principles; the second refers to the need to make compromises based on these principles; and the third demonstrates the importance of the *plurality of normative supports*[1]. Inspired by the work of Matthieu de Nanteuil, our article will thus detail different possible supports: the ethics of compromise, the ethics of capability, and the ethics of recognition (de Nanteuil, 2016). Secondly, we will plunge into the reality of the agroecological transition by looking at three concrete cases: practices of purchase-resale in the agroecological production of vegetables in Wallonia, an action-research project with an organizations of sheep farmers and veterinarians in the Millavois area of France and the management of the global plant health within a network of artisanal vegetable seed producers. We will close with a short conclusion on agroecological justice.

2. The agroecological transition as an open-ended and non-relativist process

2.1. An open-ended process, oriented towards social transformation

In the debates on the meaning of "transition," several authors contrast the notion of "transformation" with that of "transition." Andy

---

[1] To speak of "normative" or "normativity" involves that human action is structured by several rules or principles. These principles often include values, thus shaping *ethical norms*. But this is not always the case: there are also economic, political or legal norms. The idea of "normative support" is crucial in our contribution: it means that referring human action to ethical norms is a *possibility* given to the agroecological actors – not an obligation. It implies that these actors clarify the type of norm they need and explicitly mobilize them as a guide for action. In our perspective, normativity is not given in advance: it's the result of a global process, for which we formulate a *methodology*.

Stirling (2015) points to the potentially apolitical nature of "transition" by emphasizing that, in the face of environmental urgency, the watchword "transition" leads dominant actors to view public deliberation and citizen participation as a luxury that society can no longer afford. In contrast, the notion of "transformation" implies a much more open-ended approach to the problems at hand, capable of questioning established power relationships and privileges. The dynamics of transformation are thus the result of unexpected political choices, but also of less visible paths and more ambiguous pathways than those envisaged by the established order. They seek more radical changes on a large scale and over the long term, they position themselves in the progressive and radical posture described by Shattuck and Holtz-Gimenez (2011). We can oppose the *transition* towards sustainable agricultural intensification with a path of *transformation* towards agroecological agriculture (Levidow, 2018)[2]. Our article deliberately adheres to this transformative perspective, while still giving it a particular reorientation. Though we see agroecological transitions as processes of an open-ended and indeterminate nature, they nevertheless rely on normative supports that need to be grasped and made explicit. Our work has therefore consisted in bringing to the foreground what was, from a normative point of view, in the background.

2.2. Normative supports of the agroecological transition

Our theoretical question is the ethical dimension of a transformative transition. Indeed, the normativity we refer to here concerns the conditions of a just transition and, more broadly, the meaning that actors give to their decisions when they intend to pursue an ideal of social justice. We begin our reflection with conflicts of value – ethical dilemmas – that these actors encounter in their practices. Indeed, a reductive reading of the transition might suggest that in order for this transition to take place it would be sufficient to adhere to the values of the emerging model, or to oppose those that underly agricultural productivism. The reality is more complex: in practice, many values clash with each other, and actors do not have the means to make choices based on adequate normative supports.

---

[2] In order not to weigh down our text, we will use the generic term "agroecological transition with transformative ambitions."

This is why, at the interface of Marx and Weber, it seems to us more appropriate to approach the ethical question through *conflict*. Following the path opened by McIntyre (1984), such an approach renounces making "virtue" the criterion for determining "the good life." Building on the work of Lukes (1991), it focuses on the practical contexts in which moral questioning arises. Finally, and especially, such an approach seeks to take seriously the ethical experience itself, that is, the questions, divisions or indecisions that actors regularly face in the professional sphere. On a theoretical level, such an orientation has an important consequence: it leads to an acceptance of pluralism, not only of the "ideals of a good life," as said Paul Ricoeur, but of the normative supports themselves. Clearly, there exist *several* possible ways of overcoming the conflicts of value that actors face. However, this calls for two clarifications: this approach implies going beyond the simple observation of a "irreducible pluralism of values," highlighted by the founding works of John Rawls (1971); it also implies moving away from a universalizing perspective of social ethics, in favour of a more contextual approach (Hunyadi, 2012).

By the same token, the question of the relationship to action appears as the central – and no longer secondary – precept of ethical reflection. The question is no longer how to apply general, decontextualized considerations to concrete situations, but how to take the latter as starting points in identifying the normative supports available to actors in singular contexts. Let us thus take a closer look at the question of the normativity of the agroecological transition. While this is not a new question, our contribution focuses on the status of this normativity for guiding the transition. To do so, we propose an analysis in three stages: principles, compromises and plurality of normative supports.

2.3. From a statement of general principles to a plurality of models of justice: The three "normative stages" of the agroecological transition

Assuming the perspective of a transformative agroecology (Mendez et al., 2013), the Belgian Interdisciplinary Research Group on Agroecology (GIRAF) began defining in 2011 a series of principles for the transition towards sustainable food systems (Stassart et al., 2012). For the 9 co-authors (founding members of GIRAF) the aim was to define first and foremost a framework that could clearly define what agroecology was or was not, according to them. This framework is composed of 12 principles. In order to not reduce them to a juxtaposition of good practices, the group enriched Altieri's well-known principles, elaborating a series of principles touching on methodological and socioeconomic dimensions of agroecology. In this way, GIRAF laid the foundations for a *first normative stage*, to which other authors would contribute (Nicholls and Altieri, 2016).

Antoinette Dumont (2016) went on to propose a *second stage*. She endeavoured to show that the above principles do not sufficiently take into account the conflicts of value agroecology actors are confronted with on a daily basis. Her research highlights the following paradox: on a daily basis, transition actors are obliged to *negotiate* with their ideal… if they want this ideal to be *translated* into reality. Clearly, the pursuit of an effective transition presupposes the construction of compromises, in order to allow actors to get out of untenable situations. And these compromises sometimes imply depending on the productivist model, without forgetting the transformative aim of agroecology. A change of perspective thus takes place – this is the second normative stage: the question is no longer whether compromises are necessary – they are – but according to what philosophy should they be implemented in the service of a just transition. By basing her investigations on the "cities (*cités*) model" (Boltanski and Thévenot, 2006), the researcher has endeavoured to recognize these compromises, but also to examine the conditions that make it possible to turn them into a normative support.[3] The ethics of compromise thus appeared as a way of giving an ethical framework to actors' experiences. Obeying certain rules, the agreement on an intermediate solution makes it possible to maintain the plurality of values over time, in a relatively stabilized form, in the name of the common good. Nevertheless, the question remains: is this ethics of compromise the only possible way forward?

This question leads us to suggest a third stage: that which consists in making the plurality of normative supports a distinct dimension of the analysis, but also of the research framework. Compromise is then only one of the possible configurations among a plurality of normative resources. In his book *Rendre justice au travail* (de Nanteuil, 2016), Matthieu de Nanteuil identifies four possible normative perspectives: the

---

[3] Following these authors, compromise is opposed to arrangement – an unstable and informal negotiation, similar to "barter," that does not lead to a lasting commitment on the part of the protagonists.

ethics of discussion (Habermas, 1991), the ethics of compromise (Boltanski and Thévenot, 2006), the ethics of capabilities (Sen, 1999; Nussbaum, 2011) and the ethics of recognition (Butler, 1990; Honneth, 1991, 2016)[4]. The argument is as follows: if the philosophical controversy concerns which normative support has the most solid argument, the sociological question is of a different order. For social actors, it is a question of knowing which is the most appropriate support for their situation. This presupposes keeping open a range of possibilities, or never shutting it, in the place of those who are actually faced with practical dilemmas. Such an approach therefore promotes a *plural and contextual approach to social ethics*: rather than seeking to apply abstract reasoning to local situations, it is a matter of starting from the difficulties encountered by actors in order to envision *with them* the normative supports they might need to overcome these difficulties. In this perspective, the plurality of normative supports is not a "bonus" to ethical reflection, but the very condition of its effectiveness.

3. The agroecological transition in the face of the plurality of normative supports: Ethics of compromise, ethics of capability, ethics of recognition

The three cases in which we develop our reflection are situated in areas of experimentation of the agroecological transition. They are the result of doctoral research conducted by three of the co-authors of this chapter:
- the analysis of labour and workforce employment in market gardening with an agroecological transition perspective (Dumont, 2017);
- action-research with an association of sheep farmers and extension agents involved in an agroecological transition (Lacombe, 2018);
- management of global plant health in a collaboration with a network of independent vegetable seed producers (Klaedtke, 2017).

These research projects have all, to varying degrees, developed a dimension that is both transdisciplinary and transformative (Herrero

---

[4] In our case studies, the ethic of discussion did not appear to be a framework for justice in the face of the dilemmas encountered. Nevertheless, this does not make it a framework to be excluded in the context of the ecological transition.

et al., 2018), either of the course of the research (Lacombe), in the interpretation of results (Klaedtke), or in the restitution to stakeholders (Dumont). In this respect, our chapter has a specific intention: to better grasp the role of a research framework that links researchers and actors through reciprocal learning trajectories, in the emergence and stabilization of normative resources necessary for the transition. With this, we touch on the "engaged" dimension of these three research projects. As Bell and Bellon remind us (Bell and Bellon, 2018), it is not possible to dissociate the knowledge produced from what is important to us and the hopes we maintain.

The question of ethical dilemmas emerged in the course of several thesis defences (April 2017 – December 2018) in an iterative process between the reflexive work of doctoral students around the approach deployed in and around the theses and the categorization of M. de Nanteuil in his previous work. Based on an initial problematisation and the hypotheses formulated for each case, we thus have from a methodological point of view:

1. Developed for each case study an initial analysis of the dilemmas and how to overcome them, written by the doctoral student who conducted the study, in order to validate the plurality of normative supports in their respective theoretical framework.
2. The six authors then collectively built a complete analytical frame that makes it possible to compare the 3 case studies. This methodology is based on six steps: (i) identification of the tensions that surround professional practices; (ii) identification of a significant ethical dilemma; (iii) analysis of what actors consider to be a significant injustice; (iv) mobilization of a particular normative support to overcome these difficulties; (v) identification of the epistemological frameworks that this support presupposes; and (vi) stabilization of this support as a resource for transformation, within the researcher-actor relationship.
3. Finally, we completed our problematisation by resituating the issue of ethical dilemmas within the larger question of social justice as applied to ecological issues.

3.1. The dilemma of purchase-resale and the ethics of compromise

This first case study is based on the PhD work of Antoinette M. Dumont that examines working conditions in vegetable production, of organic farmers based on less than 10 hectares that commercialize their produce in "direct sale" combined for a minority with "purchase-resale." Though these farmers do not identify explicitly with agroecology, they pursue nonetheless several agroecological principles in terms of both ecological and socioeconomic factors (Dumont and Baret, 2017).

Tensions

The farmers of the two examined production systems, confronted with socioeconomic and political constraints that made the viability of their farms challenging, do not manage to put into place *all* of the principles of agroecology. As such, they violate their own ideals and the societal expectations placed on them. Although farmers do not necessarily explicitly refer to agroecology, their difficulty could be translated by the affirmation: "I am unable to implement agroecology." After many years of work, they feel judged. They are disappointed by the harmful situations that they have created to the point that, on occasion, they quit farming altogether.

A significant ethical dilemma

Behind this tension hides a profound dilemma: either undertake only direct sale commercialization on behalf of the transition, but assuming precarious working conditions; or, on the contrary, improve working conditions by undertaking a purchase-resale arrangement, but at the risk of evading certain agroecological principles. Indeed, farmers must work a considerable number of hours in order to generate a low income and find it difficult to pay their workforce properly. Some chose thus to engage in what is commonly called "purchase-resale." They generate more than 50% of their revenue by buying organic vegetables from abroad or from Walloon farms qualified as "industrial organic," bought at low prices through wholesalers and which they subsequently resell for profit. This practice is very controversial, and the vast majority of producers reject it because they do not want to depend on or support unsustainable farms. A minority group of farmers, however, have opted for this option. They see it as a necessary choice in order to live decently and assure quality employment for their labour force. They have chosen to favour values of social equity at the expense of values of autonomy and financial independence from the dominant system. The majority of producers, on the contrary, make the opposite choice. This situation points to a deeper polarization of between those who refuse any linkage with the productivist model, even if it means impinging on the viability of their farming goals, and those who seek to develop an agroecological model that is viable in the long term, for both producers and farm workers.

A significant injustice: Constant but ignored compromises

In practice, all producers must make compromises. The figure of radical injustice that they refer to, explicitly or otherwise, is unilateralism, that is, the impossibility to combine different values. This occurs when consumers, other farmers, or managers of agricultural institutions, who generally have poor knowledge of agroecological realities, judge harshly the compromises made and, more generally, produce arbitrary judgements. The feeling of a radical injustice is experienced by agroecological producers, who often feel isolated in the face of the compromises made; it can be found in the already mentioned assertion: "I am unable to implement agroecology." This affirmation translates both the considerable efforts that they make and the divisions that they experience, as a societal issue intertwines with their individual farming goals. With no way out, this phrase can be a source of suffering.

A specific normative support: Reconstructing legitimate action through the ethics of compromise

In her thesis, Antoinette M. Dumont proposes a theoretical framework called the *justification of practices* to analyse the extent to which agroecological principles guide producers' practices and affect their work experience (Dumont, 2017). This framework is based on the notions of "justification" (Boltanski and Thévenot, 1991) and the ethics of compromise (de Nanteuil, 2016).

To get out of the above mentioned dilemma, producers have implicitly created two forms of compromise. Minority producers opting for purchase-resale espouse a vision of agroecological agriculture that is based on a compromise between the "industrial" city (economic efficiency) and the "civic" city (decent work contracts, focused on employment quality), while producers opting rather for a direct sales model espouse a compromise between the "domestic" city (interpersonal relations and direct sales, "localness") and the "civic" city (transparency of the relation to the consumer, a refusal to support unsustainable practices for society as a whole)[5]. These choices govern the entire operation of their farm, materialized by different investments and contracts (for example, a predominance of workers hired with permanent, as opposed to short term, contracts). What is important to observe though, is that the minority producers also consider their action as the result of a compromise, thus intending to give a solid ethical foundation to their own choices.

Moreover, these compromises are justified by a plurality of axiological registers, including the civic city that implies a strong sense of the common good. In other words, these compromises are not purely utilitarian, nor simple superficial arrangements, but definitively *compromises* that are hard to reverse and are considered to be desirable for society.

The underlying epistemological framework

The compromise is revealed here thanks to the heuristic strength of the "cities model" (Boltanski and Thévenot, 2006). Examining the justifications that underly agroecological practices reveals the existence of not just one, but many. The question thus shifts to the coexistence of these different justifications. This is where the ethics of compromise emerges: ensuring that this coexistence takes hold in the long term, and that the transition is never a simple matter of unilateral or arbitrary decisions. The importance lies in the fact that compromises are made between actors who hold values with the same degree of legitimacy. As such, the mobilization of the "cities model" allows for researchers and actors on the ground to open to a new question: what forms of agroecology and what underlying values do we not only want to support, but to sustain?

The stabilization of normative support within the researcher-actor relationship

The researcher's conceptualization and explanation, which updates and stabilizes the normative support of the ethics of compromise, becomes a resource when actors can use it to overcome taboos and discuss

---

[5] The terms "industrial city," "domestic city" and "civic city" are borrowed from Boltanski and Thévenot (2006).

"unspoken" dilemmas. So, after the thesis, several restitutions, as well as the constitution of a file co-written with actors on the ground, became the opportunity to explain this resource and to debate the experienced dilemma, as well as the two compromises made in order to overcome it. While the themes of purchase-resale or off-the-books employment were up to that point taboo (Pongo, 2017), they are now brought up and sometimes discussed, namely within commercialization cooperatives that bring together organic producers on small or medium sized farms.

3.2. Act collectively or accompany singularities? The ethics of capabilities in the face of a recurring dilemma

The model of compromise between "cities" presupposes a willingness for negotiation more so than creating one. In the following case, the normative support used refers to an ethics of capabilities. It is based on the development of capacities for action. This figure is illustrated by the case of AVEM, an Association of Veterinarians and of around 60 Millavois Sheep Farmers. In the context of the restructuration of the Roquefort Confederation, which sets the prices and volumes of sheep milk produced for a majority of livestock farmers, and faced with reoccurring climatic uncertainty, this organization decided to put into place a project of agroecological transition towards autonomy. Some of the farmers of the steering committee, accompanied by a veterinarian and agronomist of the association, proposed to co-construct locally a diagnostic tool to assess farm's possibilities to achieve agroecology.

Tensions

The first part of their project[6] is focused on the conception of a tool for evaluating the agroecological performance of production systems. Tensions emerged in the debates surrounding the criteria and indicators to put into place. Indeed, it was necessary to identify concrete elements to which the group of farmers and partners wished to give importance and collectively attribute value. The farmers who initiated the project had

---

[6] The SALSA project (Agroecological Milk Systems of Southern Aveyron) is a Ministry of Agriculture State funded project lasting three years (2014–2017), within the framework of the "Collective Mobilization for Agroecology" call for projects.

emphasized the importance of farm autonomy, the impact of practices on the environment and the efficiency in the use of inputs. Nonetheless, during the first interviews and workshops, it became clear that for other farmers, other criteria appeared to be essential (namely, the revenue and well-being of the farmer). Moreover, the criteria to be given most importance in strategic choices differed from one individual to another.

A significant ethical dilemma

With this tool, the dilemma for the AVEM was the following: on the one hand, recognize and support the singularity of individual pathways of change; and on the other hand, to demonstrate, thanks to the evaluation of production systems, that the most autonomous farms and those least impacted by the environment were also those that were doing the best financially. Ought that all the members could be convinced of the advantages of their transition, and that the veterinary advisors and agronomists could then disseminate these best practices. But the question of individual choices and pathways of change, as well as the values that motivate individuals to change, had not been addressed directly at the start of the project. Moreover, the objective for the association was to accompany its members towards an autonomy in regard to chemical inputs as well as a decision-making autonomy with regard to other commercial actors (upstream and downstream), while helping them to minimize their environmental impact. Finally, there was an implicit desire that working on a more global approach to farms would enable a more collective dynamic, by re-articulating individual monitoring (veterinary monitoring) with more collective aspects (trainings).

A radical injustice: Actors faced with the inability to act

The initiators of the SALSA project hoped that area farmers would *be able to reflect and decide for themselves* on the quantities of milk to be produced on their farms, thanks to a broader reflection on the autonomy and coherence of their system. They noted, indeed, that more and more farms are expanding and increasing their livestock to be able to produce more milk, and thus meet the ever-increasing demand from the industry. With the end of the federal Roquefort system, which defined production references for each farm and set agreements on prices, their fear was that this phenomenon would increase, with falling prices as in the dairy cow sector. The desire was therefore to counter this phenomenon by allowing farmers to avoid having their litre-amounts dictated by downstream actors and to build their own references locally. The aim of the project and the diagnostic tool was therefore to provide farmers with the capacity to establish the litre-amounts sold to the dairy industry according to their own production, and not the other way around. The injustice they felt they were facing was the inability to act – and to act independently.

A specific normative support: Overcoming conflicts by developing the capacity of actors

The individual/collective dilemma was resolved by three shifts that have increased the actors' capacity for action. First, the choice of the form of the results produced by the tool that was able to represent, on the same graph, the results of the collective and the particularities of each farmers' situation (Fig. 1). Indeed, this graph shows a form of ideal to be attained for the group of farmers (the most autonomous and efficient farms possible, with little impact on the environment, at the top right), while at the same time making it possible to compare individual situations and envision individual pathways of change, taking into account each person's starting off points and subsequent choices.

The results were presented on the same graph to facilitate an overall comparison of systems and to facilitate exchanges between farmers. A negative autonomy corresponds to a situation where the farmer buys all his production as well as part of the feed necessary to maintain his non-lactating ewes.

Subsequently, the tool evolved in its uses. Rather than using it as a tool for *prescribing* changes, the researchers proposed to use it as a base for *demonstration and debate* within small groups of farmers. This made it possible to support individuals in their transition choices, while collectively constructing common principles of action through debate and experimentation. Finally, the tool was seen as an evolving, non-stabilized, resource within the technical committee.

The underlying epistemological framework

Camille Lacombe mobilized the conceptual framework of John Dewey's American pragmatism to take action and to analyse how the tool was used in concrete situations. Rather than deploying it solely for positivist uses, as a base to produce knowledge and identify good practices, she proposed to use it in a more constructivist way, as a support to facilitate reflective debates among farmers on the ends and means of

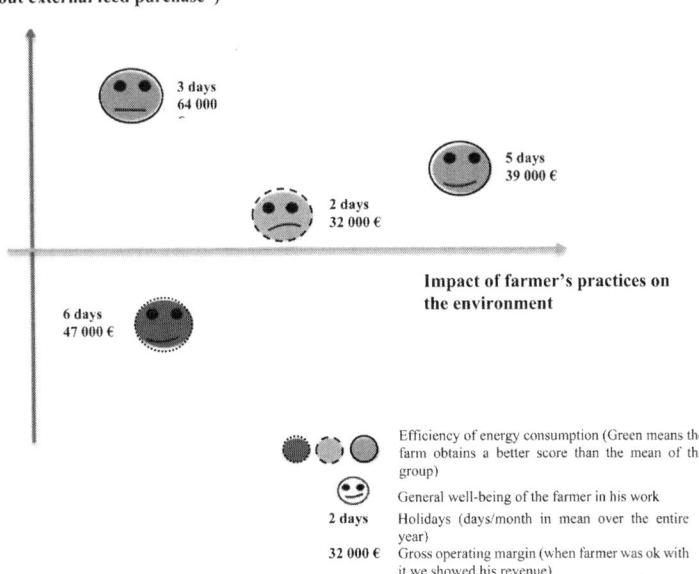

Fig. 1: Example of results of the SALSA diagnostic for four participating farms.

the agroecological transition (Lacombe et al., 2018). Pragmatism puts individuals, their choices and experiences, at the centre of reflections. It recognizes individuals as autonomous beings, capable of making choices and testing hypotheses in their work situations, in order to collectively assess the consequences. It has brought to the forefront the question of the capacity of farmers to carry out a certain number of actions or decisions. We see here how pragmatism, as a theoretical referent and epistemological framework, has made it possible to open the question of the ethical issue of developing the capacities of the AVEM.

The stabilization of normative support within the researcher-actor relationship

On this occasion, the co-produced resource is the result of a change of perspective, in which the diagnostic tool was co-constructed between researchers and the SALSA project committee. Supplemented by "socio-economic" and "well-being" dimensions, it served as a diagnostic tool and a means of comparing individual situations. It has become the basis for *explaining* individual choices and trajectories of farmers, for *discussion* among farmers about the objectives and means of transition, and for *simulating* a change in practices among farmers engaged in a transition on their farms.

The AVEM also used the results to test new methods of joint intervention between veterinarians and agronomists on farms, in order to provide comprehensive support to farmers. Participating farmers were also put in the role of advisor to their willing colleagues on changes in practice, which they discussed with the group. Finally, the AVEM took up the issue of transition support again to discuss the follow-up to the SALSA project within the association. The steering committee decided that the facilitation of the collective workdays for transition accompaniment should continue beyond the project.

Finally, the progress made within the project has made it possible to move from designing a *tool* to designing a *support system* using the tool as a basis for facilitation. This made sense for an association bringing together farmers and their advisors, where the question of how the advisors mobilize the tool with farmers ultimately proved to be as important as the question of the type of knowledge that the tool was able to produce.

3.3. An identity that breaks with European phytosanitary regulations: The dilemma of farmers' seed and the ethics of recognition

In this case, it is the ethics of recognition that will be mobilized. More than the search for middle-ground or the development of capacities, this normative support is based on the recognition of practices and a new identity: that of artisanal seed producers.

The case discussed here concerns a recent reclamation for the recognition of a paradigm shift in the treatment of plant health, established beginning in 2010 by an association of eight seed craftsmen, the "carrot crunchers" (*Croqueurs de Carottes*), a member of the "Farmers Seed Network" (*Réseau des Semences Paysannes*). These small seed companies promote the production and dissemination of open-pollinated organic vegetable seeds free of property rights. Their reclamation is based on their own plant health management practices, which break with a European phytosanitary regulatory system that follows a logic of removal of pathogenic organisms in order to eliminate the sanitary risk linked to seeds.

Tensions

Conflicting tensions emerge around the European phytosanitary regulatory constraints. During the final symposium of the European research project "Farm Seed Opportunities," an expert in seed technology stated that the sanitary quality of seeds supplied by these artisans was "mediocre" by European standards. He based this on analyses of bean seeds showing the presence of a *Xanthomonas* bacterium (*Xanthomonas axonopodis* pv. *phaseoli* and *X. fuscans* pv. *Fuscans*). This bacterium causes a disease called "common bacterial blight" on this plant. Some of the *Croqueurs* present were indignant. On the one hand, according to their experience, the disease in question is endemic: it can no longer be eradicated as the regulations aim for. On the other hand, their practices and observations lead them to live "with" the disease rather than "against" it, considering the overall health of the plant as it evolves in a given terroir rather than through the microbiological properties of the seeds alone.

A significant ethical dilemma

Following this symposium, Stephanie Klaedtke undertook a PhD project with the aim of gaining a better understanding of the management of bean health by the *Croqueurs*, using the *Xanthomonas* issue as a model. However, as it is a quarantine organism, the detection of *Xanthomonas* on seeds during trials must be reported to the authorities, potentially leading to the exclusion of the artisans seeds from market sale. For artisanal producers, the dilemma is either to enter into an open conflict in order to bring recognition to a perceived injustice, or to accept that their practices are tolerated in the margins, in order to develop both their market access and their alliance with the scientific world, but at the risk of becoming invisible.

A significant injustice: Contempt

The injustice denounced by the *Croqueurs* stems from the declaration of their supposed incompetence in seed health management by the seed technology expert (van der Brug, 2010), which appears as a form of contempt. This expert asserted that "farmers awareness of seed quality is limited and that knowledge about seed treatment is practically non-existent." This public expression of disdain is part of a broader disqualification of the practices and knowledge developed by seed artisans, particularly in regards to their claim to contribute to cultivated biodiversity[7]. For them, this situation is not tolerable. On an issue with as much importance as biodiversity, it deliberately ignores the knowledge and skills of these seed producers, accumulated over the course many years.

A specific normative support: Extending the conditions for recognition to non-institutional statutes and practices

Stephanie Klaedtke thus chose to substitute a second disease for the *Xanthomonas* model, that of "halo blight," caused by a bacterium of the genus *Pseudomonas* and that has an infectious cycle and symptoms similar to *Xanthomonas*. Both bean diseases are managed in the same way by the *Croqueurs* but *Pseudomonas* is not a quarantine organism, unlike *Xanthomonas*. Studying the interactions of beans with *Pseudomonas* does not endanger either the research project or the seed producers.

We refer to this second disease, *Pseudomonas*, as "diplomatic," because it can bring about the coexistence of different and contradictory practices (Stengers, 2006). Switching to *Pseudomonas* transforms the problem and frees the *Croqueurs* from regulatory quarantine threats. This transformation enables change of perspective within the project. By following the practices of artisanal seed producers, the project broadens the initial bio-technical perspective by focusing on the way in which stakeholders define and conceive plant health, through their approach that is both *global* in relation to the plant and *situated* in relation to the terroir.

The underlying epistemological framework

Departing from this tension between experts and seed artisans, Stephanie Klaedtke mobilized the sociology of controversies to complement field trials in an attempt to understand the issues and the networks

---

[7] An expert in plant breeding, Théo Van Hinten, says about them: « The in situ community is less coherent [than the ex situ]. [...] Not only the nature of [...] and the conservation methods vary, also the actors [...] have quite different perspectives. [...] They form a very diverse mixture [...]. As a result, [...] it is difficult to describe their specificities [because of] the lack of information. [...] It is not clear whether this material is available for utilization. [...] Making these components better accessible, by digitizing and translating them and connecting them via websites, will increase access to in situ diversity substantially ». (Farmer's Pride Workshop 1, January 2019, WG 1 1C – 2C, PGR user network stakeholders and Promoting and enabling use of material conserved in situ in the network).

that contribute to them. She thus enacted the shift from a bio-technical approach ("what biological interactions are at play?") to a socio-technical approach ("how is plant health defined?"). In her approach, she does not reduce the dispute to a simple conflict of interest. Following Callon (1999), she interprets it instead as a "hot debate" where: "Actors are unable to agree upon what constitutes causes or effects of the problem, nor on the knowledge necessary to solve it. Even a common definition of the problem cannot be agreed upon. In such hot debates, the involved actors propose visions for the future that are incompatible." This allows Klaedtke to express the depth of disagreement and to ultimately note that: "The participants don't agree with the definition of bean seed health that the seed technologist considers to be a given." (Klaedtke, 2017). She then shows how seeds, practices and knowledge circulate within the *Croqueurs* network by observing them closely – following the precepts of Actor Network Theory – in order to understand how they collectively manage the global health of plants that are in co-evolution with their terroirs. This reveals the specificity of their socio-material practices but also their socio-political practices (Hecquet, 2019). The latter reveals their public, but also legal, reclamation for recognition of another way of managing living things.

The stabilization of normative support within the researcher-actor relationship

Defended in 2017, the PhD on which this case study is based received an enthusiastic reception that surprised the doctoral student and her sociologist co-superviser, less for the academic appreciation of its transdisciplinary approach than for the interest it elicited from seed artisans.

Their interest related to the new resource that the thesis had become for their normative framework. The researcher demonstrates, in collaboration with these producers, the importance of reconfiguring the initial sanitary problem, by moving away from the paradigm of sanitary purity to a global and situated approach of plant health at the crossroads of the practices of seed artisans, their terroir and the seeds they maintain. (Klaedtke et al., 2018) How the problem is defined holds greater importance to artisan seed producers than academic solutions that do not address their "real" problems. Based in an academic qualification and the scientific network, the thesis becomes a resource to give credibility to producers' request for recognition, which relates to a different conception of managing living things but also to the identity of seed craftsman. This is also what paradoxically allows this thesis to be described as "transformative," despite the fact that it does not provide an immediate solution (Stassart et al., 2020).

This credibility and the collaborative dimension of the thesis contributed to the establishment within the Farmers' Seed Network of a working group on plant health. In July 2019, the group organized a meeting called "Visions of living things and plant health," in which the researcher and the actors of the thesis participated. The linchpin of this meeting was the farmer of the *Croqueurs* who had been outraged by the reductionist vision expressed by the expert. Taking the participants on a tour of his farm plots, he retraced the history of the controversy in which they had initially been described as incompetent. Starting from a profound dilemma, having gone through the rejection of a significant injustice, the actors and researchers together defined a path for overcoming their difficulties. The ethics of recognition provided them, in this case, with the support they needed to back up their analyses and their reclamation for political action.

4. Conclusion

In this chapter, we have intended to recognize the ideals of justice sought by the actors involved in the agroecological transition, while equipping them to overcome conflicts of value that are too often ignored. Looking at three case studies, we have observed that, in each situation, the normative supports used to overcome these conflicts were *different*. Yet, beyond the differences, a *common methodology* emerged. In this chapter, we have been able to identify six successive phases though which actors involved in the transition are able to identify profound dilemmas and to overcome them while referring to contextual norms of justice. This methodology is our major collective contribution: it confirms the existence of a pathway for an open-ended and non-relativist agroecological transition. One point, however, deserves to be explored in greater depth: that of the genesis of normative supports, given that they are linked to particular contexts of action. The question could be formulated as follows: what are these different contexts composed of and/or how do they contribute to the production of specific normative resources? Answering this question would entail launching a full research programme, capable of revealing the underlying dimensions of these different supports and making them accessible – or not – to the actors of the transition. The project of an agroecological justice still requires significant scientific work.

References

Bell, M., Bellon, S., 2018. Generalization without Universalization: Towards an Agroecology Theory, *Agroecology and Sustainable Food Systems*, 42, 6, 605–611. https://doi.org/10.1080/21683565.2018.1432003.

Boltanski, L., Thévenot, L., 2006. *On Justification – Economies of Worth*, Princeton, Princeton University Press.

Brug, J., 2010. *Improving Seed Production and Marketin, In WP3 Improving Seed Production and Marketing*, Report, Inra, Marseille.

Butler, J., 1990. *Gender Trouble: Feminism and the Subversion of Identity*, New York, Routledge.

Callon, M., 1999a. La sociologie peut-elle enrichir l'analyse économique des externalités? Essai sur la notion de cadrage-débordement, in Foray, D., Mairesse, J. (Eds.), *Innovations et performances – Approches interdisciplinaires*, Paris, Ed. des Hautes Etudes en Sciences Sociales, 399–431.

Cochet, H., 2015. *Comparative Agriculture*, Quae, France, and Springer, Netherlands.

de Nanteuil, M., 2016. *Rendre Justice Au Travail*, Presses Universitaires de France, Paris.

Dumont, A. M., 2017. *Analyse systémique des conditions de travail et d'emploi dans la production de légumes pour le marché du frais en région wallonne (Belgique), dans une perspective de transition agroécologique*, PhD, Université catholique de Louvain, Louvain-la-Neuve.

Dumont, A., Baret, P., 2017. Why Working Conditions Are a Key Issue of Sustainability in Agriculture? A Comparison between Agroecological, Organic and Conventional Vegetable Systems, *Journal of Rural Studies*, 56, 53–64. https://doi.org/10.1016/j.jrurstud.2017.07.007.

Dumont, A. M., Vanloqueren, G., Stassart, P. M., Baret, P. V., 2016. Clarifying the Socioeconomic Dimensions of Agroecology: Between Principles and Practices, *Agroecology and Sustainable Food Systems*, Taylor and Francis, 40, 1, 24–47. doi: https://doi.org/10.1080/21683565.2015.1089967.

Giménez, H., Shattuck, A., 2011. Food Crises, Food Regimes and Food Movements: Rumblings of Reform or Tides of Transformation? *The*

Journal of Peasant Studies, Routledge, 38, 1, 109–144. doi: https://doi.org/10.1080/03066150.2010.538578.

Habermas, J., 1991. *Moral Consciousness and Communicative Action*, Cambridge, MIT Press.

Hecquet, C., 2019. *Construction d'une demande de justice écologique : le cas des semences non-industrielles*, Université de Liège, Arlon.

Herrero, P., Dedeurwaerdere, T., Osinski, A., 2018. Design Features for Social Learning in Transformative Transdisciplinary Research', *Sustainability Science*, 0123456789. https://doi.org/10.1007/s11625-018-0641-7.

Honneth, A., 1991. *The Struggle for Recognition: The Moral Grammar of Social Conflicts*, New York, Polity Press.

Honneth, A., 2016. *Freedom's Right. The Social Foundations of Democratic Life*, New York, Columbia University Press.

Hunyadi, M., 2012. *L'homme En Contexte. Essai de Théorie Morale*, Paris, Cerf.

Klaedtke, S., 2017. *Klaedtke, S. Governance of Plant Health and Management of Crop Diversity – The Case of Common Bean Health Management among Members of the Association Croqueurs de Carotte*, PhD, Liège University.

Klaedtke, S., Mélard, F., Chable, V., Stassart, P., 2018. Les Artisans Semenciers, Les Haricots et Leurs Agents Pathogènes: La Biodiversité Cultivée et La Santé Des Plantes Au Cœur d'une Identité, *Etudes Rurales*, 202, 36–55. https://doi.org/10.4000/etudesrurales.14930.

Lacombe, C., 2018. *Approche Pragmatiste de l'accompagnement d'une Transition Agroécologique: Une Recherche Action Avec Une Association Éleveurs et Conseillers Dans Le Rayon de Roquefort*, PhD, Toulouse, Université de Toulouse.

Lacombe, C., Couix, N., Hazard, L., Gressier, E., 2018. L'accompagnement de La Transition Agroécologique : Un Objet En Construction: Retour d'expérience d'une Recherche-Action Avec Une Association d'éleveurs et de Conseillers Dans Le Sud-Aveyron, *Pour*, 234–235, 217–23. https://doi.org/10.3917/pour.234.0217.

Levidow, L., 2018. Sustainable Intensification, Agroecological Appropriation or Contestation, in Constance, D. (Ed.), *Contested Sustainability Discourses in the Agrifood System*, Abingdon, Taylor and Francis Group, 23–41. doi: https://doi.org/10.4324/9781315161297-1.

Lukes, S., 1991. *Moral, Conflict and Politics*, Oxford, Clarendon Press.

MacIntyre, A., 1984. *After Virtue. A Study in Moral Theory*, New York, Notre Dame Press.

Mendez, V. E., Christopher, M. B., Roseann, C., 2013. Agroecology as a Transdisciplinary, Participatory, and Action-Oriented Approach, *Agroecology and Sustainable Food Systems*, 37, 1, 3–18.

Nicholls, C., Altieri, M., Vazquez, L., 2016. Agroecology: Principles for the Conversion and Redesign of Farming Systems, *Journal Ecosystem and Ecography*, S5: 010. https://doi.org/10.4172/2157-7625.S5-010.

Nussbaum, M., 2011. *Creating Capabilities: The Human Development Approach*, Cambridge, MA, The Belknap Press of Harvard University Press.

Pongo, T., 2017. *Le Secteur Agro-Alimentaire: Des Controverses Aux Mobilisations. Analyses d'engagements Au Sein d'activités de Distribution et de Production*, PhD, Université Catholique Louvain, Louvain-la-Neuve.

Rawls, J., 1971. *A Theory of Justice*, Cambridge, MA, Harvard University Press.

Sen., 1999. *Development as Freedom*, New York, Oxford University Press.

Stassart, P., Adams, F., Catinaud, P., Chable, V., Klaedtke S., 2020 (submitted). La Fabrique d'une Thèse Transformative : Comment Hybrider Le Potentiel Transformatif et Légitimer Au Sein d'une Thèse, *Nature, Sciences, Sociétés*.

Stassart, P. M., Baret, P., Grégoire, J.-C., Hance, T., Mormont, M., Reheul, D., Vanloqueren, G., Visser, M., 2012. Trajectoire et potentiel de l'agroécologie, pour une transition vers des systèmes alimentaires durables, in Vandam, D., Streith, M., Nizet, J., Stassart, P. M. (Eds.), *Agroécologie, entre pratiques et sciences sociales*, Dijon, Educagri, 25–51. http://hdl.handle.net/2268/130063.

Stengers, I., 2006. *La Vierge et Le Neutrino, Des Scientifiques Dans La Tourmente*, Paris, Les empêcheurs de penser en rond.

Stirling, A., 2015. Emancipating Transformations, from Controlling « the Transition » to Culturing Plural Radical Progress, in Scoones, I., Leach, M., Newell, P. (Eds.), *The Politics of Green Transformations*, London and New York, Earthscan, 54–67.

Thinking through the lens of the other: Translocal agroecology conversations

DIVYA SHARMA AND BARBARA VAN DYCK

1. Situating "change" in grassroots agroecology movements

1.1. Movements making history

Omar Felipe Giraldo (2019) eloquently reminds us that agroecological transitions have been discussed mostly in terms of changes in institutional procedures or public policies needed for changing the agricultural and food regime. Less attention has been given to situating agroecological transitions in their historical contexts shaped unevenly by the global expansion of agroextractive industries and how social movements strategize to redefine territories through agroecology (Giraldo and Rosset, 2018). In their quest for building empowering relations and collective practices, these movements make their own history, "but not as they please" (McMichael, 2008; Mares and Alkon, 2012). In this chapter, we argue that agroecology movements' relationship with enacting change needs to be understood in their historical place-based context. Congruently it is important to have conversations across movements to identify how the globally connected agroindustrial food regimes create systemic challenges that manifest in concretely different forms.

Drawing on our respective engagements with two groups – the Kheti Virasat Mission (KVM), a group that advocates for natural farming in the north Indian state of Punjab, and the Agroecology-in-Action (AiA) network that advocates for agroecology and food justice/sovereignty in Belgium, we explore how "ontologies of change" are cultivated by agroecology social movements. As Lindahl (2015: 163) highlights, the characterization of (constituent) power as the "capacity to change a normative state of affairs" cannot do without an ontology. In conceptions of social change, ontology has to do "with 'what world' is, is desired or is to be made" (Moulaert and Van Dyck, 2013: 466). Thus, "ontologies of change" refer to assumptions about the realities of social life and how they can be transformed. As we will argue, accounts of existing food systems and the historical material conditions that shape them along with visions of potential and desired "becoming" are essential in shaping ontological relationships to change. Visions of desired change and the practices for constructing alternative possibilities are, as will be illustrated, necessarily connected to *what is*.

Thus, social movement struggles for agroecological transformations linked to enacting food sovereignty, continuously articulate an understanding of what has to change (what is) and the process and practices of the "not-yet-become." This exemplifies a belief in what philosopher Ernst Bloch (1959/1986) refers to as the world being always unfinished, unclosed. In that sense, transformations for enacting food sovereignty are not about filling the gap between the food systems that exist and a new preconceived agroecological one to be, but practices that continuously shape the desired change (Dinerstein and Deneulin, 2012).

Agroecology struggles challenge encroachments by extractive agroindustrial food systems to carve out spaces for enacting alternative practices (Giraldo and Rosset, 2018). In this process, a constant dialectical negotiation between defining visions for a different future and necessary transformations of the present, act as guiding beams for the performance of ongoing mobilization practices. These mobilization practices in turn reshape the boundaries of visions for agroecological transformations. The genesis of the views of *what is* and *what is desired* therefore have to be approached as social processes that never reach completion.

Movements do not enact practices mechanistically translated from predetermined visions, rather they engender political consciousness through collective thinking/action, and expand the horizon of what is imaginable and achievable in a specific context. While movements are often associated with their unified agendas and discourses employed to engage with powerful institutional actors, it is critical to be attentive to the differences within movements that reflect the historically situated lived experiences of participants positioned differentially within social hierarchies and their uneven capacities to enact transformative practices (Wolford, 2010).

The engagement of both of us in two differently positioned struggles for agroecological transformations is our starting point to disentangle what these movements' situated discourses, foci of action, and modes of organizing for food sovereignty say about how these social movements relate to change. The key insight that has emerged for us through this conversation is that even as movements are often compelled to define the firm principles of what constitutes progressive agroecological transformations to resist the industrial food system, sustained recognition of the voices and challenges of marginalized actors within localized geographies that may find it difficult to enact such principles is paramount for working towards building food sovereignty. Further, building alliances between disparate social groups and connecting across domains compartmentalized by the industrial food system and broader modernist discourse such as health, labour, ecological degradation, culturally and historically specific social hierarchies, is germane to the work of prefigurative agroecological struggles. In doing so, KVM and AiA engage in boundary work across epistemic and political divides within their regional and national contexts and contribute to creating "thick legitimacy" for agroecological transformations (de Wit and Iles, 2016). We view our translocal conversation in this chapter through these two struggles located across the Global North and the Global South as a modest attempt at such boundary work as well, that can contribute to creating "thick legitimacy" by making visible and confronting exclusions within agroecological movements in order to build solidarities.

1.2. Building solidarity in difference

The uneven but connected historical trajectories of agroindustrial expansion in particular places make it imperative that movements "start where they are" (Gibson-Graham, 2006). "Starting where you are" often entails negotiating tensions between meeting the short-term needs particularly of those most marginalized and powerless within the current food system and long-term goals of defining and enacting food sovereignty through agroecological practices. As Edelman et al. (2014) suggest the "tolerance for pluralism" and fostering transitional efforts towards agroecology is one of the biggest challenges for the food sovereignty movement (Holt-Giménez and Shattuck, 2011).

Agroecology movements have been criticized for being implicated in cultural conservatism and for their failure to challenge intersecting racialised, gendered and caste-based social hierarchies (Guthman, 2004; Khadse et al., 2017). As resistance to and delinking from state and corporate capital becomes the primary focus, addressing inequities that arise through locally manifest social hierarchies can be deprioritized. For instance, the notion of traditional farming knowledge is often a reference to physical labour-intensive techniques such as mixed cropping, irrigation apt for local ecosystems, cultivation of so-called coarse grains and foods that were a part of local diets. However, traditional farming can be invoked uncritically without an account of how these labour practices can be embedded in oppressive historically produced power relations and enacted by marginalized groups (Gregory et al., 2017). Similarly, a focus on fair prices for farmers may divert attention away from low-income food consumers, oppressive labour conditions or low wages for farm workers.

From the standpoint of food sovereignty struggles in their regionally specific forms, Edelman et al. (2014) highlight several issues that illustrate the challenges of defining the boundaries of political agendas for resisting agroindustrial food systems, while practicing open-ended socially inclusive deliberations on the meanings of food "sovereignty" and "justice" and the paths for getting there. These include thorny issues such as the relationship between interests of small producers and low-income net consumers, rural-urban dynamics and food access in ecologically deficient regions (Tornaghi and Dehaene, 2019). These tensions necessitate thinking about how agroecological production practices are situated within globally connected trading and pricing regimes and the co-ordinated strategies needed for engagement with national political regimes and international institutions to avoid shifting the adverse costs and consequences from one set of marginalized groups and/or region to another.

Even as the discourses and practices of differentiated place-based struggles are articulated in relation to the socio-ecological and political constraints imposed by agroindustrial expansion, they are not confined by its terms completely (Escobar, 2008). Competing discourses and divergent understandings within the movement on the one hand can foreclose radical transformations (Di Masso et al., 2014), but they can also show the potential for building solidarity and addressing multiple and intersecting forms of exploitation. The deterministic normative ideological visions for transformation set by a few relatively privileged actors to project unified resistance can be strategically salient within national and global institutional fora. But dialogues within and across movements are imperative for making the tensions visible, and to negotiate practices of change that do not reproduce the exclusions of agroindustrial food systems.

Based on our work as researcher-participants, we engage in such a translocal dialogue in which we first highlight how the specific histories and geographies of these agroecology movements shape their political discourse, the core issues around which they organize and critically the issues and practices that linger on the margins. Then, we will briefly explore how the core issues of the two movements may speak to each other. The conversation emerged from discussions at an INRA writing workshop (Agroecological transition, between determinist and and open-ended visions, Le Pradel, October 2018) and a brief letter correspondence to deepen reflexivity and facilitate exchange on the authors respective queries with regards to aspects of the agroecology movements in their respective contexts (see Fig. 1). We do not position ourselves as representatives of the movements we engage with or speak on their behalf but as researcher-activists with a partial reading of movement dynamics who are in solidarity with the agenda of the food sovereignty movement and the objective of making it more socially inclusive.

```
Dear Barbara,

On your question about how I ended up working with KVM...

I was motivated to work in Punjab after having worked in the
neighbouring mountainous region of Uttarakhand which
presented none of the challenges mentioned. Farming was by
default 'organic' in Uttarakhand with most people (mostly
women) growing food for subsistence consumption on very
small, terraced farms. Men migrated seasonally to towns for
work which generated cash incomes for the household. With
the implicit comparison between these two contrasting
regions in my mind, I was interested in understanding the
politics of sustainability in a region like Punjab where
agriculture was not peripheral but at the core of people's
livelihoods and identities. The unevenness between
industrialised/extractive agriculture landscapes like that
of Punjab and other rural regions peripheral to the national
development project that are visibly 'backward' in terms of
lacking infrastructure such as roads, agricultural extension
investments and marketing avenues has also manifest itself
in a pathological form.

Following workers from KVM and getting involved in their
everyday practices of work, for me, was about understanding
how they organise — whether they appeal to revival or
constructing something new, and what were the resources they
were drawing on materially (seeds, crops, plants, animals)
and socio-cultural (connections, relations, knowledges,
perceptions and memories)? Over the years, since 2012, I
also traced the improvisations and adaptations KVM made to
their organising strategies, the negotiations between
various members which included medium and large farmers,
local journalists, doctors, urban professionals and paid
workers who were mostly from landless families and did bulk
of the grassroots organising work. The paid workers were
driven by commitment to the 'cause' but also had compulsions
to generate their own livelihoods from the movement, and
therefore better understood the economic compulsions of
farmers they were organising.

Yours,
Divya
```

```
Dear Divya,

I can't wait to hear more about your work in Punjab.

Trained as a forest engineer, and worked as a researcher in
the university with neoclassical economists, I have become
increasingly frustrated with academic researchers working in
the interest of dominant political and economic powers and
their role in transforming political issues into technical
questions. With that understanding, I have become more
involved in advocacy work for food sovereignty and
agroecology.

The privatization of life through biotechnology in
particular, has been key in my understanding of the
destructive role of universities by promoting reductionist
thinking and transforming the minds of students into experts
with the confidence of transforming all kinds of problems
such as climate change, food, waste or nature into
manageable technical questions. Little by little we learned
how we can understand a whole on the basis of the study of
separate parts of it. And then also how, with these pieces
we would be able to create live. We also learned that
scientific facts exist in isolation from politics, cultural
norms, values, the social contexts in which knowledge is
produced.

Around 2010, reclaim the fields and the European Nyéléni
network have been important in bringing together people from
different organisations and backgrounds around seeds and
farmers autonomy from big AG. Over the years, agroecology
became important as a bridging concept between developmental
organisations, popular education movements, peasant and food
organisations. While a number of people in the movement (try
to) make a living from food production and transformation,
many people are based in the capital Brussels and work in
associations facilitating things such as food baskets,
access to land, and do advocacy work around the right to
food and agricultural policy.

More soon,
Barbara
```

Fig. 1: Letter correspondence as part of a translocal conversation.

2. The Kheti Virasat Mission[1]

2.1. Agroecological organizing in an epicentre of the Green Revolution

The northern state of Punjab has been an epicentre of the agricultural modernization project, the so-called Green Revolution (GR) since the 1960s in postcolonial India. The intensity of extractive agriculture in Punjab is reflected in the quantity of wheat and rice it contributes to the national food stocks despite its relatively small area and population size. With only 1.54% of the total geographical area of country Punjab has contributed 35–40% of rice and 40–75% of wheat to the central pool in the past two decades (Singh et al., 2012). It is now widely accepted that monocultures of rice and wheat have led to degraded soils and declining and contaminated groundwater. The productivity level of wheat and rice has reached a plateau, and farmers have to use higher quantities of inputs to maintain current levels of yields. Unsurprisingly Punjab has the highest per hectare usage of pesticides in the country (S.P. et al 2017). A narrow pool of hybrid wheat, rice and cotton varieties have displaced many of the other crops that were grown in the region such as pulses, millets and mustard. The groundwork for the GR was laid by the British colonial state that built an extensive canal infrastructure, enacted measures for land consolidation, and reshaped the landscape through expansion of wheat and cotton for export (Bhattacharya, 2019). The volatility of prices entrapped cultivators in debt and has generated periodic political conflict within the region (Shiva, 1989; Mukherjee, 2005).

As further agricultural intensification has become unfeasible with rising costs of cultivation, chronic indebtedness and frequent crop failures (Jodhka, 2006; Padhi, 2012), a local counter-movement has developed over the past 10 years (Sharma, 2017; Brown, 2018). *Kheti Virasat Mission* (henceforth KVM) that roughly translates as "a mission for reviving farming heritage", a registered NGO since 2005 which self-identifies as a movement was precipitated by a sense of crisis and disillusionment with statist interventions, and advocates for restructuring everyday practices of production and consumption, specifically enacting a shift toward natural farming (Brown, 2018). Since the early 2000s, largely due to the

---

[1] This analysis is based on embedded ethnographic research by one of the authors' with the movement between 2013–2015 (Sharma, 2017).

efforts of KVM and some other civil society organizations, high incidence of cancer and reproductive health issues are recognized in the public and political domain as being related to extractive chemical-intensive agriculture.

While the unsustainability of ongoing intensification is recognized by farmers, politicians and public agricultural university and extension officials, the way forward is contested. The dominant policy narrative continues to emphasize moving people out of agriculture; consolidation of landholdings to enable contract farming with direct control and involvement of corporations in selection of crops and volumes to improve efficiency; judicious/precision use of water and agrochemicals and shift to high value crops such as fruits and vegetables to move away from water guzzling rice varieties. KVM, in contrast, espouses a different vision and argues for chemical-free agriculture, reviving farmers' autonomy through regaining control over seeds, (re)generating lost knowledges and practices, establishing farmer-led marketing cooperatives and cultivating direct contact with urban consumers to provide healthy food. However, in their everyday mobilizing work KVM workers are confronted with challenges inscribed by the GR on the material and institutional landscape – dead soils that need more and more fertilizer every year; frequent pest attacks particularly in cotton; farmers caught in an unending debt cycle that limits their capacity for bearing short-term losses that come with transitioning to chemical-free agriculture; high medical expenditure on private healthcare; dependency of farmers on piecemeal solutions provided by agrochemical dealers and extension workers for crop protection and limited marketing/procurement infrastructure for crops other than wheat, rice and cotton. KVM workers also confront socio-cultural challenges inscribed and reinforced by the GR. Stigma and "backwardness" are associated with labour-intensive farming practices that are considered degrading work and have been systematically devalued and allocated to marginalized groups. For instance, dealing with cow dung for organic manure is a task that has been historically relegated to oppressed Dalit castes and women.

2.2. Shifts (and rifts) in KVM's discourse and practices

The political discourse of KVM is shaped by the critical role played by Punjab in building aggregate national food security as part of postcolonial state building. KVM's mobilization discourse, thus, emphasizes regional exploitation of natural resources by the federal state, along with socio-ecological degradation such as adverse health impacts, declining and contaminated ground water and economic losses generated by the technological treadmill for farmers. KVM also explicitly articulates a critique of postcolonial state's adoption of Western agricultural science and public policy which continues to perpetuate colonial modes of governance.

In its formative years, the movement was dominated by older medium and large landowning farmers. They, along with the founder of the movement, favoured a purist approach to natural farming – that is elimination of synthetic agrochemicals, resurrecting diversified cropping systems, creating seed banks and usage of natural inputs gathered from (non-market) sources that are generated on the farm. As their work gained traction over the years, they generated more funding which enabled the employment of a few paid workers some of whom come from the relatively low income, land poor and low caste backgrounds since the 2010s. Feedback from KVM's grassroots workers based on their experience of organizing suggested that a large majority of people did not respond to calls for shifting to natural farming. The participation of farmers in taking up natural farming practices was low after attendance at preliminary meetings in new villages. This feedback led to a shift in strategy. It was clear that invoking a purist natural farming strategy was not successful in expanding the outreach of the movement to landless workers, marginal and small farmers or even medium farmers if their households were solely dependent on agricultural incomes and caught in debt cycles. Natural farming justified in ecological terms only had resonance among a small minority of older farmers who had an ideological commitment to the movement and were supported by other household members earning secure salaries through nonfarm employment.

The shift towards a mobilizing discourse framed in economic terms as well as emphasizing health outcomes has occurred through deliberative negotiations. During village meetings, KVM workers argue that lower synthetic inputs means lower costs for farmers. They advocate a gradual reduction in chemical usage over several seasons as opposed to calls for elimination and initially changing practices on a small plot to grow food for household consumption. In addition, KVM workers focussed on engaging women who had stopped working on their family farms with mechanized agrochemical intensification. Withdrawing women from farm work had become symbolic of upward mobility and status in landowning communities (Padhi, 2012). To reverse this process, KVM workers began organizing women to cultivate vegetable gardens in their household compounds since they were reluctant to go to the farms which are generally outside the village in keeping with gendered social norms. This strategy has also allowed landless households to become involved as most households' compounds have small patches of land. Participation of women from landless households is however limited given the paucity of time as they perform waged farm labour. Trainings are conducted by women activists on how to grow seasonal vegetables and maintain these plots without using chemicals. The vegetables are primarily used for household consumption and surpluses are given or sold to others in the neighbourhood. Activists envision that women from landowning households may convince men to move towards natural farming as well. However, a majority of the farmers involved with the movement do not see themselves shifting their entire farms to organic production. They see natural farming as a way to generate healthy food for their families while they continue to grow chemical food for the market. High yields of wheat and rice are critical for generating cash incomes that they need for paying for their children's privatized education and household medical expenses. Some KVM workers argue that it is morally legitimate to produce healthy food for their own families while selling chemical laced food in the market until the state creates better incentives and infrastructure to support organic farmers. They point to the long standing exploitation of Punjabi farmers through unremunerative prices and pollution of the regional ecology by the federal state motivated by keeping food prices low for consumers.

The older purist farmers involved with the movement since inception perceive this more economistic strategy as a dilution of the agroecological agenda. For them, inclusion of farmers and rural workers who are not ideologically committed to ecologically regenerative natural farming marginalizes the goal of moving toward a diversified cropping system and constructing autonomy for farmers. These largely dominant caste farmers argue that KVM should instead focus on creating ideal diversified demonstration farms and work with committed farmers even if the movement remains confined to a small group.

On the other hand, women and landless worker activists in KVM conceive the success of the movement not in terms of creating ideal natural farming plots but in engaging people from all strata and accommodating their differential capacities. These differential ontologies of change expressed within the movement reflect the intersections of gendered, caste and class subjectivities produced within this regional ecology shaped by extractive agriculture. With the broadening of the movement's constituency the exclusions of the purist agroecological vision centred on natural farming and autonomy have been challenged, advancing the process of enacting change through and for deepening deliberative democracy.

3. Agroecology-in-Action[2]

3.1. Agroecology movement building in an urbanized region

In the very different setting of a densely populated and highly urbanized European country, Agroecology-in-Action (AiA) in Belgium has been organizing around agroecology since 2016. Since the late 1960s, agriculture in Belgium has been intensely industrializing with a rapid decrease in the number of farms, while remaining farms are larger and increasingly specialized. Labour costs as well as urbanization pressure on farm land are high. The number of land workers has dropped steadily over the last decades. Cultivated land is mainly used for grassland, cereals, corn for fodder, potatoes, sugar beets and a few more crops to feed an export-oriented livestock and processed vegetable industry. While agriculture gradually disappeared from the daily lives of most people, a series of food safety scandals and agrifood related crises (including the dioxin-contaminated feed scandal (1999), the mad cow disease (late 1990s) or the swine fever (1980s, 2018), ongoing problems of nitrate pollution, crises in the dairy sector (2010, 2018), the quick building up of massive potato surplus and plunging of prices during the COVID19 crisis (2020)) recurrently exposes the unsustainability of a highly industrialized and export dependent sector.

Whereas these scandals often frame farmers and slaughterhouses in a negative light, they reflect profound structural changes of agriculture and its increased dependency on a global food and beverage industry (Stassart et al., 2018). The incorporation of Belgian farmers in global

---

[2] This analysis is based on one of the author's work in the AiA coordination team since 2016.

value chains through cheap imported fodder, an extensive logistical infrastructure, trade agreements and the profit obligations of suppliers and retail make them highly vulnerable to fluctuating world markets. Not surprisingly, in the margins of this industrialized agrifood system, initiatives of community supported agriculture, participatory breeding networks (Baltazar et al., 2016), food box schemes (Manganelli and Moulaert, 2018), and proposals for enhancing access to affordable quality food for all (Damhuis et al., 2017) are emerging. While these initiatives and networks are small in many ways, they are increasing in number and visibility (Pleyers, 2017).

In this context, AiA was born out of an NGO-driven informal peasant farming support network which reached out to federations organizing against poverty, environmental organizations, mutual health-insurance providers, educators and researchers to meet up during a two-day agroecology forum. The call was received enthusiastically and led to what is today close cooperation among around 35 organizations. While individual member organizations have different foci ranging from advocacy work to peasant organizing, from the provision of social services to popular education, or from organizing alternative food infrastructures to research, as a network, AiA focuses on supporting and voicing food sovereignty and agroecology initiatives on a mostly regional level.

On an international level AiA is associated with the Nyéléni food sovereignty network. This transnational social movement is a network of alliances working together to enhance existing food sovereignty initiatives and to strengthen their work at local, national, regional and global levels. In 2016, AiA developed its constituting text upon the 6 principles of the Nyéléni Europe declaration (2011) – Changing how food is produced and consumed; Changing how food is distributed; Valuing and improving work and social conditions in food and agriculture systems; Reclaiming the right to the Commons; Changing public policies governing food and agricultural systems. This 'declaration of engagement' (AiA, 2016) was intended as a common frame of reference. Each principle unfolded into a number of concrete intentions for future action. The value-laden and strategic principles are not interpreted as rules, but as guidelines to put agroecology into action.

3.2. Weaving alliances through different ontologies of change

In a context ridden by food scandals, strategies for agroecology and food sovereignty translated into a quest for autonomy and democracy. Active peasant and citizens' engagement in campaigns for the liberation of seeds from patents and the struggle against genetically modified seeds exposed corporate control of food systems and agricultural science. A growing interest by a new generation of farmers and their difficulty in accessing land, put land commoning high on the agenda of young urban activists. The establishment of a cooperative for the collective buying of farmland is but one example of the initiatives where peasants and citizens have been joining forces to develop agroecology infrastructure. Moreover, the involvement of Belgian international solidarity NGOs in the local agroecology platform is also the reflection of a strategic shift in the last decade. Larger international solidarity NGOs such as Oxfam increasingly included transition of European farming systems and solidarity with European farmers as crucial for the support of farmers in other regions. The latter being squeezed through dumping of subsidized European overproduction in the global market (Godfrey, 2002).

Agroecology in the Belgian social movement context, and in contrast with the established organic sector, was thus not so much a predefined model to strive for (see the introduction to this book). Instead, agroecology operated as a bridging concept for struggles converging around issues of agrifood system transformations, including farmers' autonomy, access to land, social justice, international solidarity, the right to quality food for all (Stassart et al., 2018). For the organizations and individuals taking leadership in AiA, the assumption is that it is the laborious task of coalition building in itself that creates the collective capacity to change the state of affairs. Inspired by the alter-globalization movements (Ainger et al., 2003), NGO-based activists created connections and bridges between compartmentalized issues. The tactic of building convergence among social struggles had to both sow the seeds of other possible food systems as well as to unite forces in the advocacy for integrated food policies and food democracy. As a broad alliance, AiA focuses on the facilitation of agroecology-enabling environments.[3] This would

---

[3] Our understanding of agroecology-enabling environment is grounded in Tornaghi's conceptual work on the "food disabling city" (Tornaghi, 2017) and the

include pointing to elements in existing policies and infrastructure (subsidies, education and research) that prevents necessary change as well as the demands for support to agroecology infrastructure, including cooperatives and local food networks, and facilitating access to land for (young) farmers. With specific attention to internal democracy, sociocratic decision-making (Dauby, 2019), and the organization of a number of decentralized discussion fora, the movement has been successful in presenting unified advocacy messages at specific moments, such as during electoral campaigns or the COVID19 crisis. At the time of writing, AiA delegations, reflecting the diversity of the network, are part of collective advocacy work for the relocalisation of food systems.

The recognition of disagreement and difference in understandings, definitions and approaches to change, is a shared starting point among AiA member organizations. The procedures for membership adherence and a mandatory commitment statement have been put in place to prevent unreconcilable tensions between diverging agroecology visions. This approach to transition is grounded in the trust that building coalitions and horizontal decision-making structures is a continuous process necessary for facilitating desirable/progressive change. Such open-ended radical perspective for agroecology transition, allows the coexistence of sometimes overlapping but nevertheless different agroecology visions. The smaller peasant organization and a box scheme organization for example are devoted to the creation of alternative practices, including the establishment of localized "participatory guarantee systems" between eaters and producers. For these organizations, the multiplication of and experimentation with prefigurative practices are the foundations for building a different world. Their conceptions of change are strongly rooted in everyday life and the creation of convivial spaces (Pleyers, 2017).

The more purist interpretations of multiplying autonomous agroecology spaces of activists and peasant pioneers are met with the creation of spaces for exchange, and coalition building for food advocacy work, as preferred mechanisms for change. The peasant support coalition's broadening out to anti-poverty and health-work members in particular does not only widen the scope of privileged issues but also complexifies analysis and changes strategies through the deepening of deliberative food

---

theory of action for agroecology-enabling public policies of Gonzalez de Molina (2015).

democracy. Member organizations have different positions on the role of governments, the working class, peasants and consumers as drivers for change. Urban anti-poverty organizations, consumer driven networks and different producer networks advance vastly contrasting opinions on the agency of consumers and consequently strategies such as advocacy for "fair pricing" as a driver in agroecology transitions.

The focus on prefigurative praxis of the smaller peasant organization stands in contrast with the involvement of a larger peasant organization of mostly livestock farmers, for whom agroecology movement work is also a means to enhance transformation in the sector towards farmer autonomy. Similarly, AiA adherence to foster transformation *within* the health sector is observed for the large health organizations. For these "classic" membership-based organizations, organized around economic sectors and political affinity, being part of the agroecology network enhances the legitimacy of employees driving ecologisation agendas within the organization. For example, AiA co-convened a series of discussions during the socialist health organization's festival in the summer of 2018, including on gender and agriculture or on justice and migration, as a means to experiment with new topics and different ways of working with members. In addition to popular education, these larger organizations focus on political-economic reforms as drivers for change.

For some member organizations of AiA, change happens primarily through policy reform, while for others it happens through radical changes in experiments around cultures of living and organizing differently. As a network, AiA embodies the differences in organizing for change between "old" social movements and the "hope movements" (Dinerstein and Deneulin, 2012) with a strong focus on democratic collective decision-making processes.

4. Discussion: Muddling out of extractivist agriculture

In this section we discuss a number of connected insights generated through reflections on ontologies of change articulated by these two agroecology struggles in physically distant places.

First, both KVM and AiA agroecology movements emerged in regions with heavily industrialized farming systems. An industrialized farming ecology, however, in post-colonial India and in Western Europe are materially and socially very different landscapes (Gupta, 1998).

Whereas agriculture is peripheral in the minds of most people in Belgium, it is at the core of people's livelihoods and identities in Punjab. Agroecological transition in Punjab requires altering deeply entrenched practices promoted by current institutional infrastructure and public policy. Alternative ways of farming, and the social life associated with it live in the memories of older people but are largely invisible on the material landscape. KVM activists confront not only the economic and ecological unviability of the current farming model but also subjectivities of younger generations shaped by decades of developmentalist discourse that denounces agriculture as "backward" and moving to non-farm jobs as the desired aspirational goal. In Belgium, AiA emerged from the observation that alternative practices do exist but would benefit from pro-active alliance building in a context where peasant farming practices are politically marginalized and where food movements are mostly urban and elitist (Lagasse, 2017). A productive tension drives AiA's work. How to support the multiplication of existing socially and ecologically alternatives while also prioritizing the removal of systemic obstacles to the right to healthy food and to the very survival of peasant farming?

Second, the modes of organizing in both contexts arise from their present realities – degraded soils, farmers who have lost control over production processes and lack of access to healthy and affordable food for most – as well as from their specific desired futures (Dinerstein, 2016). The inequality between investments in agricultural extension, research and marketing infrastructure for extractive agroindustry and for the emergent agroecology infrastructure is striking in both cases. KVM and AiA organize with the aim of building components of their desired futures in the present as a prefiguration of radical agroecology praxis (Tornaghi and Dehaene, 2019). However, this is only possible in conjunction with organizing simultaneously to resist locked-in pathways of extractivist agriculture and food systems through transforming the material (access/control/use of seeds/land, water) and immaterial (ideology and ideas) realities (Giraldo and Rosset, 2018). Even as KVM actively challenges the agricultural scientific community on issues such as plant breeding and toxic agrochemicals, they do not engage with struggles for land redistribution/access struggles which would antagonize landowning farmers in the movement. Similarly, diverging positions in the network, shapes AiA's prudent engagement with issues such as meat consumption or consumer food pricing. Thus, producing malleable boundaries of political discourse is an ongoing dialectical process between effective resistance tactics against powerful actors in the agroindustrial food system and diverse practices that are enacted and negotiated by differently positioned actors within the movement. These negotiations have led to incremental shifts in in both movements' strategies, but they are also conversations about determinations, that inform the non-negotiable boundaries of its political discourse and the directionality of the movement.

Third, we found that visions of change are linked to subjectivities produced through socio-economic hierarchies in each regional ecology. Practices of organizing are accompanied by conflicts not just with powerful institutional actors, but are also contingent on the intersection of inequalities manifest within the movements' core constituencies and their immediate material environments. These multi-layered conflicts reinforce the "ontological significance of change" (Lindahl, 2015). Class inequality and urban/rural divides are reflected in AiA's transition imaginaries and strategies. The key role of urban environmental activists and small farm organizations in AiA risks the sustained reproduction of exploitative labour conditions due to marginal attention to workers' concerns in agroecology transition. Caste, class and gendered inequalities as well as generational differences are significant in shaping the negotiations in KVM around visions and practices of change. While movements' may strategically underemphasize internal conflicts and hierarchies to construct effective political narratives targeted at hegemonic actors, conflicts in the realm of everyday organizing compel the recognition of the interests and voices of most marginalized groups such as those of landless workers within the food system. These voices open the possibilities for a more inclusive food sovereignty agenda and reveal the connections between the compartmentalized domains of health, work, livelihoods and ecology within public policy that are also often mirrored in social struggles.

Finally, while the movements' goals of seeing and expanding "the possible" into a post-extractivist agriculture (the becoming) comes from their histories and lived actuality, they share a prefigurative mode of organizing. A prefigurative praxis that articulates a critique of the current development paradigm with the construction of alternatives incrementally from within (Escobar, 2008; Dinerstein, 2016). AiA and KVM primarily focus on transformation of farming and food practices in and through communities while demanding supportive institutional and public policy changes. Grassroots agroecologies as the "continuously being-makeable" enactment of food sovereignty stands in stark contrast to control-based ontologies of change. The latter conceive plotting of linear stages of transition based on narrowly framed problems in terms of productivity and performance. Control-based ontologies of change are visible within framings of sustainable agriculture that propose singular and universalized prescriptions such as use of precision technologies without recognizing the situatedness of historical geographies and actors and the necessity of "starting from where you are" (Gibson-Graham, 2006; Stirling, 2015).

5. Conclusion

To conclude, this chapter sketches our understanding of how two situated agroecology movements articulate their ontologies of change that are determined by their histories. By juxtaposing these two struggles, we, as authors involved in distant places, aim to show the importance of translocal conversations across movements for articulating and deepening an inclusive food sovereignty politics (Mares and Alkon, 2011). The multiplication of translocal conversations of this kind, lay "the ground for seeing such movements as part of the unity-in-difference," and building a base for solidarity (Sharma and Ajl, *forthcoming*). Translocal conversations draw attention to the margins and blindspots of movements, the differentiated ways in which industrial agriculture shapes regional ecologies, and therefore understandings of "what is" and "what could be". They can also foreground critical questions one struggle raises for the other.

Thinking the (quasi) urban AiA movement through the lens of KVM in rural Punjab draws attention to the invisibility of farm workers, those without access to land and voice in decision-making about organization of production that their labour makes possible in both agroindustrial and agroecological farming. More generally, concerns of farmers and rural workers in the global south compels questioning the ways in which urban consumers in the global north define strategies for the necessary reduction of the overseas land on which they depend to sustain their food systems. The ongoing discussion at AiA between environmentalists and peasant organizations about collective strategies for reducing protein dependence enabled through imported transgenic soy sustaining the Belgian meat industry is but one example. Reflecting on KVM's discourse and practices in relation to AiA, the absence of urban poor who rely on cheap food in their discourse and enactment of agroecology becomes visible. While caste, class and gendered hierarchies have become a part of negotiations in KVM with expansion of their membership to rural landless workers, their urban alliances are confined to mobilizing middle class consumers who can support farmers by buying chemical free food at premium prices. In contrast, AiA deliberately mobilizes the intersections between production and consumption and builds rural-urban connections that include low income food consumer organizations. Health interestingly acts as an important axis in both contexts to build bridges across producers and consumers by drawing attention to the relationship between toxic agrochemicals and paucity of affordable nutritious food.

Translocal conversations enable recognition of historical inequalities with regards to who benefits and pays the price of industrial "cheap food," which also shape the forms of agroecological transitions (Patel and Moore, 2017). Such recognition is crucial to anchor agroecology struggles in a global solidarity perspective where ecological reparations can be fostered and enacted. What may appear as localized issues are reflections of food systems built on a connected history of inequality, racialized human labour and colonial plantation agriculture that continue to be reproduced. Thinking through the lens of the other – translocal movement conversations – can open up agroecology movements to questions that caution against reproduction or displacement of these inequalities elsewhere or onto other marginalized groups, and broaden the space for building solidarity and reparations. In contexts of rising parochial nationalism and the co-optation of the ecologisation of farming for conservatism we believe that few things may be more urgent.

References

AiA, 2016. Déclaration d'engagement, *AiA Forum*, Brussels, Dec 9–10. [online] URL: http://www.agroecologyinaction.be/spip.php?article24&lang=fr.

Ainger, K., ed., 2003. *We Are Everywhere: The Irresistible Rise of Global Anticapitalism*, NY/London, Verso.

Baltazar, S., Visser, M., Dendoncker, N., 2016. From Seed to Bread: Co-Construction of a Cereal Seed Network in Wallonia, in IFSA (Ed.), *Proceedings 12th European IFSA Symposium*, Harper Adams University.

Bera, S., 2017. GST Rate: Fertilizers to Come Under 12% Tax Slab, Prices Likely to Rise. Livemint. URL: www.livemint.com/Politics/ KQOY1h7dQrkOdbQKrMAI2M/GST-rate-Fertilizers-to-come-12-tax-slab-prices-likely-to.html.

Bhattacharya, N., 2019. *The Great Agrarian Conquest: The Colonial Reshaping of a Rural World*, Albany NY, Suny Press.

Bloch, E., 1959/1986. *The Principle of Hope*, Volumes I, II and III, Cambridge, MA, MIT Press.

Brown, T., 2018. *Farmers, Subalterns, and Activists: Social Politics of Sustainable Agriculture in India*, Cambridge, Cambridge University Press.

Damhuis, L., Ayadi, A., Grisar, B., Grolambert, C., Hoet, J. B., Poncelet, J., Rousseau, C., Rousseau, P. M., Serré, A., Van Daele, S., 2017. Améliorer l'accès de tous à l'alimentation: Faire germer les possibles, *Rapport intermédiaire de recherche action Solenprim*, Bruxelles, FdSS.

Dauby, V., 2019. Parler d'Agroécologie, *Mémoire fin d'étude, Master interuniversitaire en agroécologie*, UG-ULB.

de Wit-Montenegro, M., Iles, A., 2016. Toward Thick Legitimacy: Creating a Web of Legitimacy for Agroecology, *Elem Sci Anth* 4. URL: https://www.elementascience.org/articles/10.12952/journal.elementa.000115/.

Di Masso, M., Rivera-Ferre, M. G., Espluga, J. L., 2014. The Transformative Agrifood Movement in Catalonia: Operational Divergences in the Construction of Food Sovereignty', Alternative Agrifood Movements: Patterns of Convergence and Divergence, *Research in Rural Sociology and Development*, 21, 159–181.

Dinerstein, C. A., 2016. Denaturalising Society: Concrete Utopia and the Prefigurative Critique of Political Economy, in Dinerstein, A. (Ed.), *Social Sciences for an Other Politics*, Cham, Palgrave Macmillan, 49--62.

Dinerstein, C. A., Deneulin, S., 2012. Hope Movements: Naming Mobilization in a Post-Development World, *Development and Change*, 43, 2, 585–602.

Dumont, A., Vanloqueren, G., Stassart, P. M., Baret, P. V., 2016. Clarifying the Socioeconomic Dimensions of Agroecology: Between Principles and Practices, *Agroecology and Sustainable Food Systems*, 40, 1, 24–47.

Edelman, M., Weis, T., Baviskar, A., Borras, J., Holt-Giménez, E., Kandiyoti, D., Wolford, W., 2014. Introduction: Critical Perspectives on Food Sovereignty, *Journal of Peasant Studies*, 41, 6, 911–931.

Escobar, A., 2008. *Territories of Difference: Place, Movements, Life, Redes*, Durham, Duke University Press.

Gibson-Graham, J. K., 2006. *A Postcapitalist Politics*, Minnesota, University of Minnesota Press.

Giraldo, O. F., 2019. *Political Ecology of Agriculture: Agroecology and Post-Development*, New York, Springer.

Giraldo, O. F., Rosset, P. M., 2018. Agroecology as a Territory in Dispute: Between Institutionality and Social Movements, *The Journal of Peasant Studies*, 45, 3, 545–564.

Godfrey, C., 2002. Stop the Dumping! How EU Agricultural Subsidies are Damaging Livelihoods in the Developing World, *Oxfam Briefing Paper*, 31.

Gonzalez de Molina, M., 2015. Agroecology and politics: On the importance of public policies in europe. In *Law and Agroecology* (pp. 395–410). Springer, Berlin, Heidelberg.

Gregory, L., Plahe, J., Cockfield, S., 2017. The Marginalisation and Resurgence of Traditional Knowledge Systems in India: Agro-Ecological 'Islands of Success' or a Wave of Change? *South Asia: Journal of South Asian Studies*, 40, 3, 582–599.

Gupta, A., 1998. *Postcolonial Developments: Agriculture in the Making of Modern India*, Durham, N.C., Duke University Press.

Guthman, J., 2014. *Agrarian Dreams: The Paradox of Organic Farming in California*, Vol. 11, Berkeley, University of California Press.

Holt-Giménez, E., Shattuck, A., 2011. Food Crises, Food Regimes and Food Movements: Rumblings of Reform or Tides of Transformation? *Journal of Peasant Studies*, 38, 1, 109–44.

Jodhka, S. S., 2006. Suicides by Farmers – Beyond 'Crises': Rethinking Contemporary Punjab Agriculture, *Economic and Political Weekly*, 41, 16, 1530.

Khadse, A., Rosset, P., Helda Morales, M., Ferguson, B. G., 2018. Taking Agroecology to Scale: The Zero Budget Natural Farming Peasant Movement in Karnataka, India, *The Journal of Peasant Studies*, 45, 1, 192–219.

Lagasse, E., 2017. Réseaux alimentaires alternatifs: élitisme ou émancipation? *Analysis Entraide & Fraternité*. URL: https://www.entraide.be/IMG/pdf/6-reseaux.pdf.

Lamine, C., Schmitt, C., Palm, J., Derbez, F., Petersen, P., (forthcoming). How Policy Instruments May Favour an Articulation Between Open Ended and Deterministic Perspectives to Support Agroecological Transitions? Insights from a Franco-Brazilian Comparison.

Lindahl, H., 2015. Possibility, Actuality, Rupture: Constituent Power and the Ontology of Change, *Constellations*, 22, 2, 163–174.

Manganelli, A., Moulaert, F., 2018. Hybrid Governance Tensions Fuelling Self-Reflexivity in Alternative Food Networks: The Case of the Brussels GASAP, *Local Environment*, 23, 8, 830–845.

Mares, T. M., Alkon, A. H., 2011. Mapping the Food Movement: Addressing Inequality and Neoliberalism, *Environment and Society*, 2, 1, 68–86.

McMichael, P., 2008. Peasants Make Their Own History, But Not Just as They Please, *Journal of Agrarian Change*, 8, 2–3, 205–228.

Moulaert, F., Van Dyck, B., 2013. Framing Social Innovation Research: A Sociology of Knowledge Perspective, in Moulaert, F., et al. (Eds.), *International Handbook on Social Innovation*, Cheltenham, Edward Elgar, 466–-480.

Mukherjee, M., 2005. *Colonializing Agriculture: The Myth of Punjab Exceptionalism*, New Delhi, Sage Publications.

Nyéléni Declaration Europe, 2011. *Food Sovereignty in Europe Now!* Krems, August 21. URL: https://viacampesina.org/en/nyeleni-europe-2011-declaration-food-sovereignty-in-europe-now/.

Padhi, R., 2012. *Those Who Did Not Die: Impact of the Agrarian Crisis on Women in Punjab*, New Delhi, SAGE.

Patel, R., Moore, J. W., 2017. *A History of the World in Seven Cheap Things: A Guide to Capitalism, Nature, and the Future of the Planet*, Berkeley, University of California Press.

Pleyers, G., 2017. The Local Food Movement in Belgium: From Prefigurative Activism to Social Innovations, *Interface: A Journal for and About Social Movements*, 9, 1, 123–39.

Sharma, D., 2017. *Techno-Politics, Agrarian Work and Resistance in Post-Green Revolution Punjab, India*, PhD Dissertation, Cornell University, Ithaca.

Sharma, D., Ajl, M. *The Green Revolution and Transversal Countermovements: Recovering Alternative Agronomic imaginaries in Tunisia and India* (under review).

Shiva, V., 1989. *The Violence of the Green Revolution: Ecological Degradation and Political Conflict in Punjab.* na.

Singh, J., Grover, D. K., Tejinder, D. K., 2012. State Agricultural Profile – Punjab, *Agro-Economic Research Centre Department of Economics and Sociology Punjab Agricultural University.* URL: http://aercpau.com/assets/docs/Punjab%20Profile.pdf.

S.P, Subash & Chand, Prem & S, Pavithra, & Balaji, S.J. & Pal, Suresh. (2017). Pesticide Use in Indian Agriculture: Trends, Market Structure and Policy Issues. Available on http://www.ncap.res.in/upload_files/policy_brief/pb43.pdf

Stassart, P., Crivits, M., Hermesse, J., Tessier, L., Van Damme, J., Dessein, J., 2018. The Generative Potential of Tensions within Belgian Agroecology, *Sustainability*, 10, 6, 2094.

Stirling, A., 2015. From controlling 'the transition' to culturing plural radical progress, in Scoones, I., Leach, M., Newell, P. (Eds.), *The Politics of Green Transformations*, Abingdon, Routledge, 54–67.

Tornaghi, C., 2017. Urban Agriculture in the Food-Disabling City: (Re) defining Urban Food Justice, Reimagining a Politics of Empowerment, *Antipode*, 49, 3, 781–801.

Tornaghi, C., Dehaene, M., 2020. The Prefigurative Power of Urban Political Agroecology: Rethinking the Urbanisms of Agroecological Transitions for Food System Transformation, *Agroecology and Sustainable Food Systems*, 44, 5, 594–610.

Wolford, W., 2010. *This Land is Ours Now: Social Mobilization and the Meanings of Land in Brazil*, Durham, N.C., Duke University Press.

The rhetorics of agroecology: Positions, trajectories, strategies

MICHAEL BELL AND STÉPHANE BELLON

1. Introduction

The word agroecology has come a long way. Since it was coined almost one century ago as an academic term, it has crossed into a wide variety of social worlds (Doré and Bellon, 2019). And although it began as an English word, translations of the term now can be found across the languages and countries. Worldwide institutions such as FAO and IFOAM now use the term, and it has also started to appear in the names of professional associations of several countries. A steadily increasingly number of degrees coin themselves as "agroecology" programs, although sometimes also referring to organic and sustainable agricultures and food systems (Nicot et al., 2018).

But is everyone talking about the same thing? One does not have to look hard to note differences in how the term agroecology is used. Is that a problem or an advantage of the term? Or is agroecology just a new open-ended catchword that is bound to eventually suffer the same fate as the word sustainability, bent this way and that until it scarcely stands for anything? Or does it need to be more determinist – more certain of how the world does and should work – in order to retain its transformative potential in the face of converging uncertainties such as energy, water, food, and climate change? Do we encourage resilience by having a more open-ended or more determinist view of what agroecology means? As the term agroecology is only just now accelerating its spread in the vocabularies of the world, this is an apt time to weigh and guide its rhetorical prospects as a transformative program for agricultural and food systems. We intend this paper as a contribution to those prospects.

Many have noticed considerable diversity in what is intended by the word agroecology. In 2018, the Agroecology Info Pool from the Swiss foundation Biovision identified 23 definitions of agroecology[1]. Some authors have suggested dichotomies between agroecologies: hard versus soft (Dalgaard et al., 2003), strong versus weak (López-i-Gelats et al., 2016), transformative versus conformative (Levidow et al., 2014), radical versus reformist (Holt-Giménez and Altieri, 2013), and scientific versus political narrative and social action (Léon, 2009), sometimes with a gradient between opposite polarities (Calame, 2016). Still others have proposed more complexities in the meanings of agroecology, with a wider range of categories (Buttel, 2003; Hubert, 2012; Brym and Reeves, 2016; Norder et al., 2016). There is nonetheless some recognizable coherence amid this diversity. On the whole, these many views seem to agree that agroecology is an integrative framework for designing alternative materialities of farming and food systems, in most instances while also valuing cultural and biological diversity and moving towards greater complexity, resilience, and social justice. They do not use the term randomly, albeit with a significant heterogeneity.

What these varying uses and categorizations suggest is that different stakeholders in agroecology have different purposes. It is as if a group of travelers – "fellow travelers," for the most part! – have come to the same station but are taking trains to different destinations. In this paper, we present an analysis of the rhetorical strategies of agroecology, which we relate to their varying purposes. Our point is not to argue that agroecology is only rhetorical, but rather to consider the narrative challenges involved in the transformation of socio-ecological relations and the creation of new agricultural "holons," as we will term them (Bland and Bell, 2007; Bell and Bellon, 2018). We will show that one of the central challenges is a tension between more open-ended and more determinist views. We then briefly apply this analysis to the framing of dynamic agroforestry systems. We conclude with a discussion of the rhetorical decline of the term "sustainable agriculture," and how those who use the term agroecology might work to prevent a similar fate for it.

---

[1] https://www.agroecology-pool.org/agroecology/definitions/

2. The rhetorical strategies of agroecology as a transformative program

We develop our analysis in three steps. First, we describe the narrative diversity of pragmatic goals of agroecology. We argue that a study of the strategies of agroecology must utilize a pragmatic perspective for understanding the goals of a narrative, which we base on Wezel et al. (2009) and their proposal to consider agroecology jointly as science, practice, and social movement. We suggest a modification of this proposal, considering other triads and including politics as a central narrative goal – a broader category than social movement – as Rivera Ferre (2018) has also suggested. Many appropriations of agroecology seem odd to consider as a social movement, such as corporate uses of the term, or overlook the issue of its valuation[2].

Second, based on this modified triangle of goals of appropriating the term agroecology, we examine the boundary strategies of agroecological narratives, moving from how appropriations of agroecology try to coordinate action and create a "holon" – a socio-ecologic "holding togetherness" of actors and action, parts and wholes, tensions and intentionalities, within varying ecologies of context. In order to hold together and coordinate action, a holon narrative must build relationships within as well as contend with external relations and their narratives, creating both a sense of connection and disconnection. Here we distinguish between what we term the *strength* of narrative boundaries, which may be relatively weak or strong, and what we term their *permeability*, which may be relatively open or closed. Importantly, narrative strength and permeability shape a holon's relations with its context. At the same time, contexts shape the direction – the sense of motivational pull – of intentionality toward a vision or plan. Intentionality emerges out of the contexts in which it seeks to act, possibly transforming them (Bland and Bell, 2007). Context and narrative are interactive, each shaping the other.

Third, we introduce a time dimension to narratives, or what we term their *trajectories*. The shaping of a narrative as more open-ended or more determinist is centrally an issue of how it considers itself within time.

---

[2] The concept of "valuation" borrowed from the economy, can be given a wider extension, referring to the modes of enhancement of practices, processes, products, etc. In contrast, some authors seek to provide evidence about the economic potential of agroecology or about its viability.

The goals and the boundaries of a narrative may change through time, either because of a deliberate strategy worked out in advance or because of contextual effects that the narrative encounters – or both. Transformation necessarily entails a trajectory, including asynchronous trajectories of biological entities, human capabilities, institutional settings, and more. But is that transformative trajectory presented as open to continual development or as more certain from the outset of its means and ends? We note especially here that narratives with relatively strong and closed boundaries tend towards determinist trajectories that do not welcome, at least initially, responsiveness to context.

2.1. Goal strategies

People do not just say anything at all about anything at all. Any speech act is an effort to accomplish an effect. Otherwise, why speak? The attempt to do something through speech we refer to as the pragmatic goals of a narrative. To summarize pragmatism in a sentence, what we say about the world depends on what we are trying to do in it, and on how the world responds to those efforts. Charles Sanders Peirce (1878) put it this way: "Consider the practical effects of the objects of your conception. Then, your conception of those effects is the whole of your conception of the object." This famous "pragmatic maxim," as it is often called, comes from a widely cited article Peirce wrote called "How to Make Our Ideas Clear." But, admittedly, it's not a very clear statement. So, we will attempt to clarify Pierce and get the point down to just two words. Here goes: Effects matter.

Wezel et al. (2009) remains the most comprehensive analysis of this link between meaning and purpose in agroecology. They suggested considering agroecology jointly as science, as practice, and as social movement, emphasizing that the three dimensions are intertwined. Conversely when considered separately, any one instance of the use of the term agroecology may be mobilized in support of sciences, practices, and movements, in infinitely varying degrees and possible combinations.

For the purpose of this paper, the semantics of these appropriations is less central. Our focus is on what appropriations of agroecology are trying to achieve, and how they mobilize the term to those ends. Consider, for example, conservation agriculture and its variants, such as Direct Seeding Mulch-based Cropping. Although often using the term agroecology, the conservation agriculture narrative program focuses primarily on soil life (AFD, 2014). Yet it is relatively unconcerned with the use of herbicides, the connections between above and below ground biodiversity, and issues of social justice and food sovereignty. Its goal is less the reconstruction of an agroecological holon than the maintenance of one through the adjustment of existing relations in the face of contextual pressures, and it rarely uses the language of social movements to describe itself. Similarly, integrated production focuses on plant protection issues on specific crops or plots within a farm, rather than a deep redesign of socio-ecologic relations (Navarrete et al., 2012) and engagement with social movements. The transformational goals of such uses are relatively modest, in contrast to other rhetorics of agroecology. In contrast, the rhetorical deployment of agroecology by organizations such SOCLA (the Latin American Scientific Society of Agroecology) and ARC (the Agroecology Research-Action Collective) gives a central place to issues of social justice, and portrays agroecology as "transforming our food and political systems towards justice and ecological health." ARC seeks to build a holon that includes a greater range of actors within food and agriculture. As ARC describes, "as engaged scholars, we commit to working within and alongside the social movements of farmers, workers, organizers, and advocates."[3] These different visions reflect different goals, and thus different politics.

As a result, we would like to modify slightly the "triangle of agroecology" that Wezel et al. (2009) readily implies, and that one of us has earlier been involved in making more explicit. There are many politics of agroecology, not only bottom-up politics, and we wish to provide a means for understanding the full sweep of the pragmatics of agroecology as a narrative strategy. The triangle of agroecology can be easily modified to suit our purposes in this paper. We need only widen the term movement out by re-labeling that corner of the triangle "politics." Another approach might be to add another category, differentiating politics from social movements. But that seems to us a secondary level of analysis. One might as well also subdivide agroecology as a science into, say, social science and natural science, or agroecology as a practice into, say, organic or non-organic practices to mention some common tensions in agroecological discussions. For our purposes here, a higher level of abstraction is clearer. In what follows, therefore, we will analyze the pragmatic goals

---

[3] https://agroecologyresearchaction.org/home/

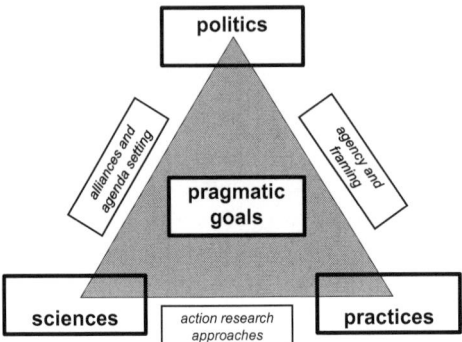

Fig. 1: A modified "triangle of agroecology", with three vertices and edges.

of agroecology as sciences, practices, and politics; all being pluralized (Fig. 1).

We see our interpretation of the three pragmatic goals as in line with Sevilla Guzman and Woodgate (2012), although they criticize the triangle of agroecology approach for not recognizing that any science also involves practices and politics. We agree that any science is very much also practices and politics. But we are analyzing narrative strategies, which frequently represent these vertices of the triangle as separate. Depending on their pragmatic goals and social locations, different narrative strategies of appropriations of agroecology may find such separations advantageous.

2.2. Boundary strategies

In addition to an analytic framework for understanding agroecology's narrative goals, we offer a framework for analyzing *boundary strategies*. Here we move from considering the pragmatic goals of various definitions of agroecology to how varying definitions seek to advance narrative goals by how they place themselves in *context* with other definitions. For example, this typically happened within the FAO global dialog, where various definitions were suggested during regional conferences (Loconto and Fouilleux, 2019), and needed to consider themselves in light of each other.

The specific formulation of the goals of a definition is certainly crucial to the strategy underpinning a definition. Definitions may both seek to stabilize a holon of actors who can connect themselves to a narrative, and also may deliberately seek to destabilize other holons by cross-cutting existing narratives, maintaining coordinated action, or trying to establish a new holonic coordination of action, both social and biophysical. This was exemplified by Rivera Ferre (2018) in a comparison among agroecology narratives based on the analysis of policy documents from different political actors (civil society, governments, and intergovernmental organizations). But whether seeking to stabilize or destabilize holonic boundaries, agroecological narratives arise in response to their ecologies of context, both social and biophysical, which compel a narrative change. In this sense, both stabilizing and destabilizing narratives represent at least a degree of novelty in holonic configuration.

In order to do so, however, a definition must provide potential actors within the new holon a narrative role, or they are unlikely to seek to coordinate with it. The definition must create connections that enable the coordination of actors' intentionalities. We propose that intentionality is the primary criterion for identifying and bounding a holon. By intentionality we mean the active envisioning and seeking of a set of goals, such as the farm household working and planning so that they may continue to derive a livelihood by collecting milk from cows (Bland and Bell, 2007). As well, they must also create disconnections that direct intentionalities and their biophysical conditions toward the new holon, and away from others toward which they might direct their activities and goals.

In short, a definition needs a boundary strategy. We suggest here two basic axes of boundary strategies in the construction of an agroecological narrative. While we apply these to the specific instance of definitions of agroecology, we consider these axes more general issues that the narrative underpinnings of a holon face. One basic axis might be termed the *strength* of a definitional boundary. Does it represent itself as considerably different from other narratives of coordinated action, and strongly reconstructive of them, entailing considerable destabilization and reorganization? Or does the definitional boundary emphasize only a relatively minor point of difference that does not alter the basic framework of the coordination of action? The former we call a *strong* boundary, and the latter we call a *weak* boundary. The other basic axis might be termed the *permeability* of a definitional boundary. Does it exclude intentionalities involved in other coordinated holons of action, perhaps even to the point of rejecting them? Or does the definitional boundary strive to create identification across and between holonic boundaries, engaging their differences rather than rejecting them? The former we call a *closed* boundary, and the latter we call an *open* boundary.

Combining these two axes yields four possibilities – ideal types, of course, with infinite gradations in between (Fig. 2). A boundary strategy may be strong and closed, strong and open, weak and closed, and weak and open. A strategy which is strong and closed is both radically different and socially oppositional. A strategy which is strong and open reaches for an engaged radicalness. The "normal science" we often encounter in the academe is an example of a boundary which is weak but closed, not challenging basic tenets but referencing a sharply defined in-group. A weak and open strategy is common to meliorism, phrased to be accommodating to existing ideas and interests.

However, there may be considerable contextual differences in how a boundary strategy is intended and how it is perceived. A boundary which is intended to be strong but open may have considerable difficulty maintaining that position, as the strong may be widely understood by those outside the boundary as also being closed. The very strength of a narrative may seem to present such a challenge that those outside the boundary find it convenient to merely label it as uncharitable and inward looking, and thus of little relevance. Similarly, a narrative which is intended to be meliorist and inoffensive may find that even a weak and open strategy is rejected so strongly that others counter-define it as both strong and closed, thereby controlling it by isolating it. Related scenarios could be sketched for each of the four boundary configurations, elaborating how an intended narrative effect is not the same as the actual narrative effect, as other holons struggling to hang together resist narrative counter movements.

To put it another way, boundary strategies are constructed both from within and from without. Therefore, we urge caution about immediately labeling any one boundary configuration as more worthy than another. For example, we suspect that the two open strategies would seem more immediately benign to readers of this paper. And, all other things being equal, we would agree that they are. But all other things generally are not equal. Strategies are always situated, especially the most successful ones. Perhaps a more transformative agroecology entails maintaining narrative

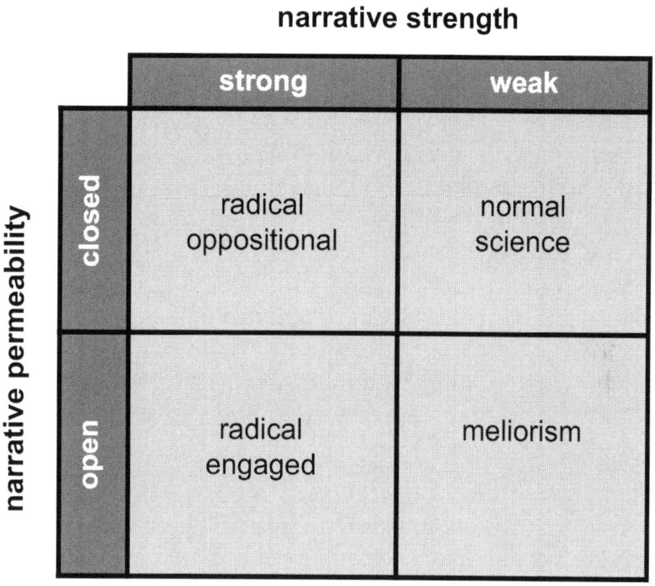

	strong	weak
closed	radical oppositional	normal science
open	radical engaged	meliorism

narrative strength (columns) / **narrative permeability** (rows)

Fig. 2: Narrative strength and narrative permeability.

boundaries which are less open – a narrative that captures rather than one which is hunted.

2.3. Time and narrative strategies

Because of their situatedness, narrative strategies also have a time dimension, in three basic ways. A narrative articulates a strategy *of time*, *across time*, and *within time*. By a strategy "of time" we mean the extent to which a narrative presents its own persistence from the outset. To articulate narrative strength and permeability is to articulate the extent to which the narrative declares a more open-ended (relatively open and weak) or determinist (relatively closed and strong) approach, and thus the extent to which it envisions that it may change as time unfolds. By a strategy "across time" we mean the extent to which a narrative, probably less explicitly, intends to move from a more determinist to a more open-ended rhetoric, or vice versa, a deliberate plan for achieving its pragmatic goals. By a strategy "within time" we mean the extent to which a narrative strategy of relative open-endedness or determinism pivots in response to the actual unfolding of events, which can never be fully anticipated, encouraging change to meet its pragmatic ends.

We refer to these three varying strategies of time as narrative *trajectories*, which themselves vary in the degree to which they are planned from the outset, or represent new configurations that actors attempt as they respond to the reactions – the "effects," in pragmatist terms – of earlier attempts. Notably, these trajectories are not mere rhetoric. They represent efforts to respond as the context they enter changes and as the holon they attempt to assemble and hold together changes as well. The same may be true of narrative goals, which may shift as the intentionalities they address reconfigure. As well, narratives may shift in order to maintain earlier goals as much as the goals themselves may change in response to context. But fundamentally, some holon must be maintained or attained. Otherwise, the narrative will echo away into nothingness.

For example, a narrative may begin with a weak and open meliorist rhetorical trajectory of time. Across time, when alliances and commitments have been secured, its trajectory may subsequently go strong and closed. Once a holon has been well established, it is better able to use its holding-together-ness as a source of power and influence over its context, and have less need of compromise. As well, the narrative commitment of constituents within the holon are investments not easily shifted, allowing a strong narrative to be less likely to burst the holon. But the best laid plans of mice and narratives go oft awry, and a subsequent pivot within time back to a weak and open narrative, or perhaps at least a strong and open narrative, may become necessary to maintain or expand a holon.

For a second example, we can consider just the reverse trajectory: a narrative of time that begins strong and closed – i.e. rather determinist – but subsequently across time goes weak and open. In this case, the strategy might be to begin by building a solid holonic core, based on those intentionalities who are ready for the narrative, who can immediately see their interests advanced by it, and who control the narrative's trajectory through their strong identification with it. Once the core is built, the plan might be to present a strong contrast that challenges other narratives to reckon with its difference, spreading recognition of the strong conception, if not agreement with it. Then, in what may initially come across as something of a surprise across time, the narrative may find success in reaching out by downplaying its differences and attempting to create identification across the narrative boundary. And if this doesn't result in successful pragmatic effects, holonic needs within time may shift the narrative trajectory yet again.

A third trajectory might be of a narrative that begins strong and closed and stays strong and closed, hoping to effect greater change of its context and a greater internal strengthening of its holon through challenge and lack of compromise. Another might begin weak and open and stay that way, so as to encourage as much buy-in from as a wide a range of intentionalities as possible. And so on, with trajectories that begin strong and open, or weak and closed, and shift as they go, as the particular case may be.

Furthermore, the trajectories of narratives could shift in the goals they emphasize, not only in their boundary strategies. For example, a narrative might begin from a scientific footing, seeking the legitimacy of science's typical mantle of neutrality. Then it might shift to practices, without coming across as an advocate of particular social positions. Finally, it might then feel that it had gained sufficient success in these two realms to present an explicitly political framing and still claim lack of bias, because of its scientific and technical accomplishments. Many academic narratives of agroecology, it seems to us, attempt precisely this trajectory. Or conversely, a narrative might follow just the opposite trajectory, moving from political advocacy of particular interests to practical techniques that support those interests and, finally, to science that might further develop those techniques and interests. In this case, the coordination of intentionalities begins with an emphasis on identity and recognition of a potential holonic entity. That identity will want practical outcomes, but may well find that it cannot engage science immediately, due to science's typical narrative of neutrality. Yet, alternative ways or webs can be identified to legitimate research in agroecology (Montenegro and Iles, 2016). The technical need for practices that support the intentionalities, however, provides a point of narrative mediation. Science can engage in the technical requirements of the practical without directly compromising its position by advocating for an identity. Both grassroots movements like Landless Workers' Movement (MST) and corporate interests like Monsanto, it seems to us, have tried to reconfigure their context through versions of this trajectory, eventually winning support of some scientists.

Other trajectories of agroecological narrative are also possible, including those that start from practices, and move from there either to politics or to science. For example, extension services might begin from an emphasis on agroecological practice in response to a context of constituent pressure, then seek scientific validation, and then feel confident enough to try to reshape structures of governmental support. The Amish might be an example of a narrative that begins from practice, then builds a political identity, and subsequently some engagement with a scientific narrative (although in the case of the Amish, the latter narrative goal remains quite weak). Trajectories are also unlikely to be linear. There may be revisions "within time," as we termed it, and even reversals, as intentionalities reckon with changes in their ecologies of contexts or discover that they had misunderstood those ecologies to begin with.

But these are not mere trajectories in what appropriations of agroecology are saying about agroecology. They are also trajectories of people and their ecologies of contexts. Each change from strong to weak, open to closed, or science to practice to politics represents a materially different kind of engagement with the world, and a materially different coordination of intentionalities and the ecology of actions they pursue. Closed narratives, whether radical oppositional or normal scientific, both constitute more deterministic accounts, and thus seek to construct flatter narrative trajectories, albeit typically entailing reconstruction in the face of countervailing narrative accounts of how matters actually unfolded.

3. Agroforestry and narrative change

With the case of agroforestry, we exemplify how narratives changed from a normal science to multifunctionality and to a policy-oriented paradigm. Agroforestry systems are varied, complex and characterized by uncertainty and sometimes, unpredictable dynamics. For a long while, they have been considered as illustrative of alternative production systems, along with other patterns such as multiple cropping, alley cropping and cover crops (Altieri, 1987). As for agroecology (Doré and Bellon, 2019), agroforestry has been redefined in keeping with different paradigms and strands (Van Noordwijk et al., 2016).

Formerly, agroforestry was devised to assist in increasing the productivity of fragile and widespread ecosystems while at the same time either rehabilitating them or arresting the process of degradation. It was then defined as a system of land management that "combines the protective characteristics of forestry with the productive attributes of both forestry and agriculture. It conserves and produces" (King, 1987). The main aim is to enhance positive interactions among trees and crops and/

or livestock, with a strong focus on biophysical components of the system (photosynthetic pathways and management of light environment) and its expression in terms of Land Equivalent Ratio (LER). LER is "the ratio of the area under sole cropping to the area under intercropping needed to give equal amounts of yield at the same management level. It is the sum of the fractions of the intercropped yields divided by the sole-crop yields" (FAO glossary). In a hypothetical scenario intercropping a grain crop with a fruit tree crop, a LER = 1.4 means that a total of 1.4 ha of sole cropping area would be required to produce the same yields as 1 ha of the intercropped system. Much attention is given to the crop yield, together with a strong forest legacy, including in modeling (e.g. Yield-SAFE model, in van Der Werf et al., 2007). When combined with "precious trees," one driver is to know whether the yield of a cash crop can be similar to the same crop in pure stand.

This first strategy we identify as having a deterministic narrative trajectory. The set of production methods applied, usually in individual plots with linear tree planting patterns, is widely determined by the expected final state of trees; for example, for having commercially valuable trees in 25 or 30 years (after planting trees or after cutting a coppice). In this case, cultivating field crops or grass in between tree rows is feasible, while ensuring that the yearly yield is comparable to a sole-crop regime. Much attention is given to early development stages of trees, namely with pruning and training, to ensure that trees have an adequate architecture for their future. It is usually implemented and managed at plot level. This strategy has then been extended to multifunctional land use, considering a wider range of goods and services. As an alternative procedure for agroforestry diagnosis and design (D&D), an open-ended approach was suggested (Raintree, 1990). Within a particular context, it fosters design creativity for large scale formal agroforestry projects (involving multidisciplinary, often multi-institutional teams of scientists).

In a second paradigm, agroforestry was defined as "a dynamic, ecologically based, natural resource management system that, through the integration of trees on farms and in the agricultural landscape, diversifies and sustains production for increased social, economic and environmental benefits for land users at all levels" (Van Noordwijk et al., 2016). The main focus is on landscape scale, including a wider number of purposes, users and arrangements among diverse patches, albeit more inspired by experiences in watershed management than agroecosystem functioning. The stance on multifunctional land use is maintained, and even epitomized with the rhetoric of environmental services and the debates on land sparing versus land sharing (Fischer et al., 2014), considering that intensification is not restricted to agriculture but can also concern forests. In spite of the beginning of the above mentioned definition, dynamics appear as secondary. Considering such dynamics can enable a second strategy, including in so-called "marginal" lands. With both planning and adaptations, vegetation management is more flexible, and vegetation states and subsequent dynamics are less predictable in terms of composition, levels of diversity, etc. Here, the local final state is not as important as the range of possibilities to obtain resources from a diversity of plant species and layers in a wider system. Diversity and diversification are key assets, with various grazing utilization patterns, multiple connections between functions and resources, reversibility and bifurcations, degrees of freedom and development of security devices (Milestad et al., 2012; Bonaudo et al., 2013; Boval et al., 2014). Time is not linear and dynamics are considered in wider entities such as agroecosystems, beyond the individual plot (Nesme et al., 2010).

A third policy-oriented agroforestry paradigm emerged as a consequence of the increasing attention given to communities living in territories and to socio-economic dimensions, beyond usual bio-physical components (Van Noordwijk et al., 2016). Institutions and policies were also at stake, especially to include agroforestry as a legitimate land use category, since agriculture and forestry were separate domains. For example, in 2013 the European Parliament defined agroforestry systems and the Common Agricultural Policy (CAP) could provide financial support for the "Establishment of agroforestry systems" (Reg. 1305/2013). In this new paradigm, a boundary work perspective on linking knowledge with action and policies is at stake (Reid et al., 2006; Kristjanson et al., 2009; Clark et al., 2011; Tomich et al., 2011). It is also articulated with Sustainable Development Goals (SDGs) giving room to new narratives such as agroforestry as institutional response to contested resource access, allowing gender and social equity enhancement and source of empowerment. It leads to a new definition of agroforestry: "a contraction of the terms agriculture and forestry, is land use that combines aspects of both, including the agricultural use of trees. This includes trees on farms and in agricultural landscapes, farming in forests and at forest margins and tree-crop production, including cocoa, coffee, rubber and oilpalm. It includes interactions between agriculture and forestry as policy domains" (Van Noordwijk et al., 2016). This broader definition extends the term to the contribution of trees to the agricultural production (including fruit production).

Such paradigms or their combinations were also found in Brazilian agroforestry case studies, identifying how agroecological practices are appropriated for diverse trajectories of territorial development (Levidow et al., 2019).

4. The narrative powers of agroecology

As agroecologists consider and compare their narrative strategies, we suggest that they keep in mind the history of the term "sustainability" (e.g. Mensah and Ricart Casadevall, 2019). This history is increasingly recognized among agroecologists as a sad story. Rather than transforming our agroecological relations, sustainability is now widely used as a narrative tool for legitimating the status quo. What could be greater evidence of the sustainability but the status quo, we often hear, for the status quo is the actually existing, and thus shows what can be sustained. Sustainability also provides an easy mantle of moralism that obscures destructive practices with references to various minor initiatives, what commentators often refer to as "green-washing."

How did this happen? How did agroecologists (and fellow travelers!) lose control of the word sustainability? We suggest looking at the narrative strategy that most uses of the term have embodied. In an effort to attract rapidly a wide following, sustainability from the start has been dominated by a weak and open boundary strategy of time that allowed many intentionalities to find a place within its story. As well, its trajectory began by emphasizing a set of practices, as in the famous Bruntland Commission definition of sustainable development as using the resources of the present in such a way as to not compromise their use in the future. From practices, the dominant sustainability narrative moved towards the sciences (Blann and Light, 2018), without forthrightly representing itself as politics, advocating particular interests. Advocates hoped that across time the move toward science would allow a strengthening of the narrative while maintaining an open boundary. But narratives and their trajectories are not completely pliable within time. Without an explicit politics, efforts to use sustainability to advance the position of the disadvantaged have made it hard to defend the term from take-over and redirection by corporate interests and the status quo. Advocates hoped that scientific narrative developments like sustainability "metrics" would give the term the strength needed to fend off such take-over. But now we must recognize that the question of what to measure and what not to measure is deeply political. We simply cannot measure everything, and many of the most important things to measure are the very hardest things to measure. Without a starting point in politics, the term sustainability rapidly became something anyone is entitled to claim and to proclaim. Sustainability's weak and permeable boundaries invited many into its narrative, but also allowed many to hollow it out.

How do we prevent a similar outcome for "agroecology"?

First of all, we advise recognizing another analytic point about narratives that we have only implied thus far: that narratives, in fact, contain both *internal* and *external* dimensions. We don't mean here the distinction narrative scholars make between meta-narrative and narrative, or how the structure of a narrative carries its point as much as its content does. We mean those within a narrative boundary of coordinated action may tell quite a different story among themselves than they tell to others outside that boundary. For example, we have little doubt that most early advocates of the term sustainability had a strong political dimension internally, for we ourselves have often heard this political narrative in college classrooms and conference barrooms, and sometimes see it in print media directed at those within the boundary. But at moments when that boundary meets administrative officials and the popular press, the politics of sustainability are almost always immediately silenced, and it becomes a set of practices documented by the sciences.

Choices of narrative strategy matter. Agroecologists clearly have effective internal narratives that gather them together into a coordination of diverse intentionalities. Most of those internal narratives, for all their diversity and contest, strike us as having a strong political dimension – and indeed, as striving to locate themselves in the middle of the triangle of agroecological narratives. But the external narrative has a simpler and more two-dimensional trajectory in which, like sustainability, politics has generally been de-emphasized. The work of Altieri has long been (Altieri 1987) a prominent exception to agroecology's generally two-dimensional external narratives, of course, as have the popular narratives of many Latin American social movements. But elsewhere, we sense a history of some panic and concern about making agroecology political. This is especially evident in the definitions of agroecology offered by academics in the early and mid 2000s (Dalgaard et al., 2003; Francis et al., 2003; Gliessman, 2007), given the apolitical quality of the larger academic narrative, which is so central to the context of these authors.

We suggest that agroecology maintain all three dimensions of the triangle in its external narratives, as well as in its internal narratives. Without a focus on political justice, we predict that the agroecological narrative will fall victim to the same take-over and hollowing of sustainability's weak and permeable strategy. Such a focus, however, does not necessarily trap agroecology in the quadrant of radical opposition, where it can be easily ignored. The radicalness of a vision does not necessarily entail impermeability. The openness and closedness of a narrative depends on how a narrative is said, and to whom it is said, not on what it says.

Indeed, recent developments in agroecological writing suggest that the context of fear of the political in the academe is starting to fade. The newest definitions of agroecology by academics have greatly elevated its political dimension. We see three main factors at work here, allowing a new direction in the trajectory of agroecology. First, academic agroecologists have had some good success in institutionalizing the agroecological narrative in the academe. They have had this success in part precisely because they did not ground the term externally in politics. Second, this success leaves the term in a moment of danger when its hitherto relatively weak and open approach has made it increasingly popular. Writers seem to recognize this danger, and are now explicitly seeking to distance the term from sustainability, and to embrace politics explicitly. Third, our understandings of science as polluted if political are fast changing to ones that see science without politics as blind if apolitical. But whether readers of this paper agree that we ought to applaud this new external narrative of explicit politics, we argue that their own narratives will be best served by attention to the dynamics of human persuasion we outline here. Maybe agroecology needs to be both open-ended and determinist to be both persuasive and transformative.

References

AFD, 2014. *Agroecology: Evaluation of 15 Years of AFD Support. Summary of Final Report*, *ExPost* n°58, GRET-AFD (Agence Française de Développement).

Altieri, M. A., 1987. *Agroecology: The Scientific Basis of Alternative Agriculture*, Westview Press, Boulder, CO.

Bell, M., Bellon, S., 2018. Generalization without Universalization: Towards an Agroecology Theory, *Agroecology and Sustainable Food Systems*, 42, 1, 1–7.

Bland, W., Bell, M., 2007. A Holon Approach to Agroecology, *International Journal of Agricultural Sustainability*, 5, 4, 280–294.

Blann, K., Light, S., 2018. Sustainability Science and "Ignorance-Based" Management for a Resilient Future, in Pimbert, M. (Ed.), *Food Sovereignty, Agroecology and Biocultural Diversity. Constructing and Contesting Knowledge*, Routledge Studies in Food, Society and the Environment, 93–114.

Bonaudo, T., Bendahan, A. B., Sabatier, R., Ryschawy, J., Bellon, S., Léger, F., Magda, D., Tichit, M., 2013. Agroecological Principles for the Redesign of Integrated Crop-Livestock Systems, *European Journal of Agronomy*, 57, 43–51.

Boval, M., Bellon, S., Alexandre, G., 2014. Agroecology and Grassland Intensification in the Caribbean, *Sustainable Agriculture Reviews*, 14, Springer Ed., 159–184.

Brym, Z., Reeve, J., 2016. Agroecological Principles from a Bibliographic Analysis of the Term Agroecology, *Sustainable Agriculture Reviews*, Springer Ed., 203–231.

Buttel, F., 2003. Envisioning the Future Development of Farming in the USA: Agroecology between Extinction and Multifunctionality? *New Directions in Agroecology Research and Education*, 1–14.

Calame, M., 2016. *Comprendre l'agroécologie. Origines, principes et politiques*, Charles Léopold Meyer Eds.

Clark, W., Tomich, T., van Noordwijk, M., et al., 2011. Boundary Work for Sustainable Development: Natural Resource Management at the CGIAR, *PNAS*. http://dx.doi.org/10.1073/pnas.0900231108.

Dalgaard, T., Hutchings, N., Porter, J., 2003. Agroecology, Scaling and Interdisciplinarity, *Agriculture, Ecosystems and Environment*, 100, 39–51.

Doré, T., Bellon, S., 2019. *Les mondes de l'agroécologie*, Enjeux Sciences, éditions Quae.

Fischer, J., Abson, D., Butsic, V., et al., 2014. Land Sparing versus Land Sharing: Moving Forward, *Conservation Letter*, 7, 3, 149–157.

Francis, C., Lieblein, G., Gliessman, S., et al., 2003. Agroecology: The Ecology of Food Systems, *Journal of Sustainable Agriculture*, 22, 99–118.

Gliessman, S. R., 2007. *Agroecology: The Ecology of Sustainable Food Systems*, 2nd Edition, CRC Press, Boca Raton, FL.

Holt-Giménez, E., Altieri, M., 2013. Agroecology, Food Sovereignty and the New Green Revolution, *Agroecology and Sustainable Food Systems*, 37, 1, 90–102.

King, K. F. S., 1987. The History of Agroforestry, in Steppler, H. A., Nair, P. K. R. (Eds.), *Agroforestry: A Decade of Development*, Nairobi, Kenya, ICRAF, 1–11.

Kristjanson, P., Reid, R., Dickson, N., et al., 2009. Linking International Agricultural Research Knowledge with Action for Sustainable Development, *PNAS* 106, 5047–5052.

Léon, T., 2009. Agroecología: desafíos de una ciencia ambiental en construcción, in Altieri, M. (Ed.), *Vertientes del pensamiento agroecológico: fundamentos y aplicaciones*, SOCLA, Medellín, Colombia, 45–67.

Levidow, L., Pimbert, M., Vanloqueren, G., 2014. Agroecological Research: Conforming – or Transforming the Dominant Agro-Food Regime? *Agroecology and Sustainable Food Systems*, 38, 10, 1127–1155.

Levidow, L., Sansolo, D., Schiavinatto, M., 2019. Agroecological Practices as Territorial Development: An Analytical Schema from Brazilian Case Studies, *Journal of Peasant Studies*. DOI: https://doi.org/10.1080/03066150.2019.1683003.

López-i-Gelats, F., di Masso, M., Binimelis, R., Rivera-Ferre, M., 2016. Agroecology, In Thompson, P. B., Kaplan, D. M. (Eds.), *Encyclopedia of Food and Agricultural Ethics*, Springer Ed., Dordrecht, NL.

Mensah, J., Ricart Casadevall, S. (Reviewing editor), 2019. Sustainable Development: Meaning, History, Principles, Pillars, and Implications for Human Action: Literature Review, *Cogent Social Sciences*, 5, 1.

Milestad, R., Dedieu, B., Darnhofer, I., Bellon, S., 2012. Farms and Farmers Facing Change – The Adaptive Approach, in Darnhofer, I., Gibbon, D., Dedieu, B. (Eds.), *Farming Systems Research in the 21st Century: The New Dynamic*, Springer Ed., Dordrecht NL, 365–385.

Montenegro de Wit, M., Iles, A., 2016. Toward Thick Legitimacy: Creating a Web of Legitimacy for Agroecology, Elementa: Science of the Anthropocene, 4, 000115.

Navarrete, M., Bellon, S., Geniaux, G., et al., 2012. L'écologisation des pratiques en arboriculture et en maraîchage. Enjeux et perspectives de recherches, *Courrier de l'environnement de l'Inra*, n°62, 57–70.

Nesme, T., Lescourret, F., Bellon, S., Habib, R., 2010. Is the Plot Concept an Obstacle in Agricultural Science? A Review Focussing on Fruit Production, *Agriculture, Ecosystems & Environment*, 138, 133–138.

Nicot, R., Bellon, S., Loconto, A. M., Ollivier, G., 2018. The European Networks of Research, Education and Training Stakeholders in Agroecology, *Open Agriculture Journal*, 3, 1, 537–552.

Peirce, C. S., 1878. How to Make Our Ideas Clear, *Popular Science Monthly*, 12, 286–302.

Raintree, J. B., 1990. Theory and Practice of Agroforestry Diagnosis and Design, in MacDicken, K. G.,Vergara, N. T. (Eds.), *Agroforestry: Classification and Management*. Wiley Interscience, New York, 58–97.

Reid, R., Tomich, T., Xu, J., et al., 2006. Linking Land-Change Science and Policy: Current Lessons and Future Integration, in Lambin, E. F., Geist, H. (Eds.), *Land-Use and Land-Cover Change: Local Processes and Global Impacts*, Springer-Verlag, Berlin, 157–171.

Rivera-Ferre, M., 2012. Framing of Agri-food Research Affects the Analysis of Food Security: The Critical Role of the Social Sciences, *International Journal of Sociology of Agriculture and Food* 19, 2, 162–175.

Rivera-Ferre, M., 2018. The Resignification Process of Agroecology: Competing Narratives from Governments, Civil Society and Intergovernmental Organizations, *Agroecology and Sustainable Food Systems*. https://doi.org/10.1080/21683565.2018.1437498.

Rosset, P., Altieri, M., 2017. *Agroecology: Science and Politics*, Fernwood Publishing and Practical Action Publishing, Canada.

Sevilla-Guzman, E., Woodgate, G., 2012. Agroecology: Foundations in Agrarian Social Thought and Sociological Theory, *Agroecology and Sustainable Food Systems*, 37, 1, 32–44.

Tomich, T., Brodt, S., Ferris, H., et al., 2011. Agroecology: A Review from a Global-Change Perspective, *The Annual Review of Environment and Resources*, 36, 193–222.

van der Werf, W., Keesman, K., Burgess P., et al., 2007. Yield-SAFE: A Parameter-Sparse Process-Based Dynamic Model for Predicting Resource Capture, Growth and Production in Agroforestry Systems, *Ecological Engineering*, 29, 4. DOI: https://doi.org/10.1016/j.ecoleng.2006.09.017.

van Noordwijk, M., Coe, R., Sinclair, F., 2016. *Central Hypotheses for the Third Agroforestry Paradigm within a Common Definition*, Working paper 233, Bogor, Indonesia: World Agroforestry Centre (ICRAF) Southeast Asia Regional Program.

Wezel, A., Bellon, S., Doré, T., Francis, C., Vallod, D., David, C., 2009. Agroecology as a Science, a Movement and a Practice, A Review, *Agronomy for Sustainable Development*, 503–515.

Postface

The INRAE Action and Transition (ACT) division has made significant investments in recent years related to a specific scientific priority called "agroecology for action," and this book is one of the major tangible developments of those efforts.

Analyzing transitional changes in agroecology has been for years an explicit research focus for transition processes within the ACT, along with other interrelated areas of investigation: the multiscale dimension of agroecological transitions (farm, food chain and territories), interactions within food system dynamics, and the way the innovation system (extension specialists, research and education stakeholders, companies) helps foster or lock in innovative experiences and support change. This book details many convincing elements of these interactions through an original framework that approaches change from two perspectives (deterministic and open-ended) and explores their possible combinations.

I would like to extend a warm thanks to the coordinators Claire Lamine, Danièle Magda (as co-leaders of the scientific priority), Terry Marsden and Marta Rivera Ferre, and to the authors for their work on this book, which explores these two ontological links to change with the purpose of clarifying contents and implications. Agroecological transition paths are complicated and lock-in issues abound, and our capacity to build a sustainable future with complex systems and considerable uncertainty is still well below our needs. Such clarifications, opportunities to reflect, and debate on the deeply rooted ways we deal with change are highly needed today. This book elucidates the collective efforts that are needed to stimulate and support substantial change in agrifood systems. The more general considerations about how we address change may also be useful in other fields.

Research dealing with our ontological links to change is highly important for "sciences for action" as they are developed within the ACT division. These combine interdisciplinary and participatory stances and model the coevolution of social and ecological/technological systems. It's also important for the biotechnical sciences which make up INRAE's core research activity and provide benchmarks, drivers and models to foster agroecological transitions.

On a more general level, we must consider the status and contributions of objective analytical knowledge, technological innovation, and crop and livestock system modelling in both deterministic and open-ended paths of change. What can agronomy and livestock farming systemic sciences contribute, and more specifically, what resources can innovative design offer to bring about change? Transition is not only about change. It is also about innovation – creating something that is desirable but which often does not yet exist. This debate is largely open and should be high on the research agenda. It should support new areas of investigation in all disciplines and especially in the ways these disciplines represent and integrate complexity, uncertainty, farming system approaches, stakeholder experiences and operational knowledge.

Of course, the implications of this book should not be limited strictly to science, but also applied to extension organizations (their missions and methodologies), education stakeholders and policymakers. I believe it is extremely important to encourage debate with these organizations and delve deeper into various topics, such as how to share knowledge on ontologies, uncertainty and complexity or how to evaluate R&D initiatives and define policies when the target is undetermined. There are many worthwhile issues still to explore.

Benoit Dedieu

Head of the INRA-SAD research division (now INRAE-ACT) (2012–2020)

ÉcoPolis

Collection fondée par Marc Mormont

Depuis sa création en 2002, la collection EcoPolis analyse les changements qui se produisent simultanément dans la société et dans l'environnement, quand celui-ci fait l'objet d'une préoccupation majeure. Au fil des volumes, la collection suit l'évolution des problématiques environnementales, en interrogeant la manière dont ces dernières sont devenues consubstantielles des questions économiques, sociales et politiques.

L'environnement a longtemps été défini comme extérieur à la société, comme un monde où la nature et les écosystèmes constituent le soubassement matériel de la vie sociale. Les politiques d'environnement avaient alors pour but de « préserver », « protéger », voire « gérer » ce qui était pensé comme une sorte d'infrastructure de nos sociétés. Après quelques décennies de politiques d'environnement, la nature et l'environnement sont devenus des objets de l'action publique et il apparaît que c'est dans un même mouvement que chaque société modèle son environnement et se construit elle-même. Cette dialectique prend une importance accrue quand les sociétés se trouvent confrontées, du fait de la montée en puissance des changements globaux, aux transformations irréversibles de l'environnement (changement climatique, perte de biodiversité, etc.). C'est le rapport entre la dialectique sociétés / environnement et ces transformations que la collection entend explorer.

Directeurs de collection :
Xavier Arnauld de Sartre et Olivier Petit

Comité scientifique

Arnauld de Sartre Xavier, géographe, Laboratoire de recherches Passages, CNRS, Université de Pau et des Pays de l'Adour, France.

Bauler Tom, économiste, *Institut de Gestion de l'Environnement et d'Aménagement du Territoire* (IGEAT), Université Libre de Bruxelles, Belgique.

Chailleux Sébastien, sociologie politique, Laboratoire de recherches Passages, Université de Pau et des Pays de l'Adour, France.

Claeys Cecilia, sociologie de l'environnement, Laboratoire Population Environnement Développement (LPED), Aix-Marseille Université, France.

Ghiotti Stéphane, géographe, centre de recherches Acteurs, Ressources et Territoires dans le Développement (ART-Dev), CNRS, Montpellier, France.

Hamman Philippe, sociologue, centre de recherches Sociétés, Acteurs et Gouvernement en Europe (SAGE), Université de Strasbourg, France.

Kull Christian, géographe, Institut de géographie et durabilité, Université de Lausanne, Suisse.

Lewis Nathalie, sociologue, Groupe de recherche interdisciplinaire sur le développement régional, de l'Est du Québec (GRIDEQ), Université du Québec à Rimouski, Canada.

Mormont Marc, sociologue, unité de Socio-Économie, Environnement et Développement (SEED), Université de Liège, Belgique.

Nahrath Stéphane, politiste, Institut de hautes études en administration publique (IDHEAP), Université de Lausanne, Suisse.

Petit Olivier, économiste, Centre Lillois d'Études et Recherches Sociologiques et Économiques (CLERSE), Université d'Artois, Arras, France.

Puerta Silva Claudia, anthropologue, Facultad de Ciencias Sociales y Humanas e Instituto de Estudios Regionales, Universidad de Antioquia, Medellin, Colombie.

Rey Valette Hélène, économiste, Center for Environmental Economics, Université de Montpellier, France.

Titre parus

N° 37 – Claire Lamine, Danièle Magda, Marta Rivera-Ferre, Terry Marsden (eds.), *Agroecological transitions, between determinist and open-ended visions*, 2021, 318 p.

N° 36 – Victor Bailly, Rémi Barbier, François-Joseph Daniel, *La prévention des déchets. Innovations sociales, action publique et transition sociotechnique*, 2021, 370 p.

N° 35 – Rhoda Fofack-Garcia, *La société des eaux cachées du Saïss. Ethnographie d'un basculement hydro-technique*, 2021, 252 P.

N° 34 – Valérie Deldrève, Jacqueline Candau, Camille Noûs (dir.), *Effort environnemental et équité. Les politiques publiques de l'eau et de la biodiversité en France*, 2021, 534 p.

N° 33 – Christine BOUISSET et Sandrine VAUCELLE (dir.), *Transition et reconfigurations des spatialités*, 2020, 348 p.

N° 32 – Jochen SOHNLE (ed./dir.), *Environmental Constitutionalism: What Impact on Legal Systems?*, 2020, 244 p.

N° 31 – Cécilia CLAEYS (ed.), *Mosquitoes management. Between environmental and health issues*, 2019, 208 p.

N° 30 – Sylvain GUYOT, *La nature, l'autre frontière. Fronts écologiques au Sud (Afrique du Sud, Argentine, Chili)*, 2017, 310 p.

N° 29 – Divya LEDUCQ, Helga-Jane SCARWELL et Patrizia INGALLINA (dir.), *Modèles de la ville durable en Asie. Utopies, circulation des pratiques, gouvernance*, 2017, 424 p.

N° 28 – Ludovic GINELLI, *Jeux de nature, natures en jeu. Des loisirs aux prises avec l'écologisation des sociétés*, 2016, 240 pages.

N° 27 – Xavier ARNAULD DE SARTRE, *Agriculture et changements globaux. Expertises globales et situations locales*, 2016, 204 pages.

N° 26 – Bernard HUBERT et Nicole MATHIEU (dir.), *Interdisciplinarités entre Natures et Sociétés. Colloque de Cerisy*, 2016, 396 pages.

N° 25 – Arnaud BUCHS, *La pénurie en eau est-elle inéluctable ? Une approche institutionnaliste de l'évolution du mode d'usage de l'eau en Espagne et au Maroc*, 2016, 331 pages.

N° 24 – Valérie DELDRÈVE, *Pour une sociologie des inégalités environnementales*, 2015, 243 pages.

N° 23 – Zhour BOUZIDI, *Se coordonner dans un périmètre irrigué public au Maroc. Contradictio in terminis ?*, 2015, 373 pages.

N° 22 – Laura SILVA-CASTAÑEDA, Étienne VERHAEGEN, Sophie CHARLIER, An ANSOMS (dir.), *Au-delà de l'accaparement. Ruptures et continuités dans l'accès aux ressources naturelles*, 2014, 244 p.

N° 21 – Xavier ARNAULD DE SARTRE, Monica CASTRO, Simon DUFOUR, Johan OSZWALD (dir.), *Political ecology des services écosystémiques*, 2014, 288 pages.

N° 20 – Céline GRANJOU, *Micropolitiques de la biodiversité. Experts et professionnels de la nature*, 2013, 202 pages.

N° 19 – Corinne LARRUE (dir.), *Le régime institutionnel d'une nouvelle ruralité. Analyses à partir des cas de la France, des Pays-Bas et de la Suisse*, 2013, 214 pages.

N° 18 – François BERTRAND et Laurence ROCHER (dir.), *Les territoires face aux changements climatiques. Une première génération d'initiatives locales*, 2013, 269 pages.

N° 17 – Véronique ANCEY, Isabelle AVELANGE, Benoît DEDIEU (dir.), *Agir en situation d'incertitude en agriculture. Regards pluridisciplinaires au Nord et au Sud*, 2013, 419 pages.

N° 16 – Cécilia CLAEYS and Marie JACQUÉ (eds.), *Environmental Democracy Facing Uncertainty*, 2012, 185 pages.

N° 15 – Josiane STOESSEL-RITZ, Maurice BLANC, Nicole MATHIEU (dir.), *Développement durable, communautés et sociétés. Dynamiques socio-anthropologiques*, 2012, 230 pages.

N° 14 – Philippe HAMMAN, Christine BLANC et Cécile FRANK, *La négociation dans les projets urbains de tramway. Éléments pour une sociologie de la « ville durable »*, 2011, 246 pages.

N° 13 – Denise VAN DAM, Michel STREITH et Jean NIZET (dir.), *L'agriculture bio en devenir. Le cas alsacien*, 2011, 140 pages.

N° 12 – Philippe HAMMAN et Jean-Yves CAUSER (dir.), *Ville, environnement et transactions démocratiques. Hommage au Professeur Maurice Blanc*, 2011, 291 pages.

N° 11 – Géraldine FROGER (dir.), *Tourisme durable dans les Suds ?*, 2010, 316 pages.

N° 10 – Muriel MAILLEFERT, Olivier PETIT et Sandrine ROUSSEAU (dir.), *Ressources, patrimoine, territoires et développement durable*, 2010, 283 pages.

N° 9 – Philippe HAMMAN et Christine BLANC, *Sociologie du développement durable urbain. Projets et stratégies métropolitaines françaises*, 2009, 260 pages.

N° 8 – François MÉLARD (dir.), *Écologisation. Objets et concepts intermédiaires*, 2008, 214 pages.

N° 7 – David AUBIN, *L'eau en partage. L'activation des règles dans les rivalités d'usages en Belgique et en Suisse*, 2007, 247 pages.

N° 6 – Géraldine FROGER (dir.), *La mondialisation contre le développement durable ?*, 2006, 315 pages.

N° 5 – Laurent MERMET (dir.), *Étudier des écologies futures. Un chantier ouvert pour les recherches prospectives environnementales*, 2005, 411 pages.

N° 4 – Jean-Baptiste NARCY, *Pour une gestion spatiale de l'eau. Comment sortir du tuyau ?*, 2004, 342 pages.

N° 3 – Pierre STASSART, *Produits fermiers : entre qualification et identité*, 2003, 424 pages.

N° 2 – Cécilia CLAEYS-MEKDADE, *Le lien politique à l'épreuve de l'environnement. Expériences camarguaises*, 2003, 245 pages.

N° 1 – Edwin ZACCAÏ, *Le développement durable. Dynamique et constitution d'un projet*, 2002 (2ᵉ tirage 2003), 358 pages.

www.peterlang.com